Kindred Brut

T0227954

Kindred Brutes

Animals in Romantic-period writing

Christine Kenyon-Jones

Routledge
Taylor & Francis Group

LONDON AND NEW YORK

First published 2001 by Ashgate Publishing

2 Park Square, Milton Park, Abingdon, Oxon OX14 4RN
711 Third Avenue, New York, NY 10017, USA

Routledge is an imprint of the Taylor & Francis Group, an informa business

First issued in paperback 2016

Copyright © 2001 Christine Kenyon-Jones

Christine Kenyon-Jones has asserted her moral right under the Copyright, Designs and Patents Act, 1988, to be identified as the author of this work.

All rights reserved. No part of this book may be reprinted or reproduced or utilised in any form or by any electronic, mechanical, or other means, now known or hereafter invented, including photocopying and recording, or in any information storage or retrieval system, without permission in writing from the publishers.

Notice:
Product or corporate names may be trademarks or registered trademarks, and are used only for identification and explanation without intent to infringe.

British Library Cataloguing in Publication Data

Kenyon-Jones, Christine
    Kindred Brutes: Animals in Romantic-Period Writing. – (The Nineteenth Century Series)
    1. English literature—18th century—History and criticism. 2. English literature—19th century—History and criticism. 3. Animals in literature. 4. Romanticism. I. Title.
    820.9'362'09034

US Library of Congress Cataloging in Publication Data

Kenyon-Jones, Christine
    Kindred Brutes: Animals in Romantic-Period Writing / Christine Kenyon-Jones.
        p.   cm. – (The Nineteenth Century Series)
    Includes bibliographical references and index.
    1. English literature—19th century—History and criticism. 2. Animals in literature.
    3. Darwin, Charles, 1809–1882—Influence. 4. Human–animal relationships in literature.
    5. Evolution (Biology) in literature. 6. Romanticism—Great Britain. I. Title. II. The Nineteenth Century Series (Aldershot, England).
    PR468.A56K46 2001
    820.9'362'09034–dc21   2001022171

ISBN 13: 978-0-7546-0332-0 (hbk)
ISBN 13: 978-1-138-25390-2 (pbk)

Contents

Illustrations

The Nineteenth Century
General Editors' Preface

The aim of this series is to reflect, develop and extend the great burgeoning of interest in the nineteenth century that has been an inevitable feature of recent decades, as that epoch has come more sharply into focus as a locus for our understanding, not only of the past, but of the contours of our modernity. Though it is dedicated principally to the publication of original monographs and symposia in literature, history, cultural analysis and associated fields, there will be a salient role for reprints of significant texts from, or about, the period. Our overarching policy is to address the widest scope in chronology, approach and range of concern. This, we believe, distinguishes our project from comparable ones, and means, for example, that in the relevant areas of scholarship we both recognise and cut innovatively across such perimeters as those designated by the designations 'Romantic' and 'Victorian'. We welcome new ideas, while valuing tradition. It is hoped that the world which predates yet so forcibly predicts and engages our own will emerge in parts, as a whole, and in the lively currents of debate and change that are so manifest an aspect of its intellectual, artistic and social landscape.

Vincent Newey
Joanne Shattock

University of Leicester

Acknowledgements

I should like to give my warm thanks and acknowledgements to the following people and organisations for their help with this book:to those who have discussed, read or commented on my work in progress, including Jonathan Bate, Anne Barton, Bernard Beatty, Clare Brant, Paul Curtis, Kirsten Daly, Trudi Darby, Siân Ede, Monica Farthing, Peter Graham, Mary Grover, Daniel Karlin, Lisa Leslie, Ruth Mead, George Myerson, Sue Tumath and Frances Wilson; to Peter Cochran for enabling me to quote from his transcriptions of Hobhouse's diary; to Virginia Murray of John Murray (Publishers) Ltd, Haidée Jackson of Newstead Abbey and Alex Alec-Smith for help in tracking down Byroniana of all sorts; to Erika Gaffney, Kirsten Weissenberg and all at Ashgate for their prompt and courteous assistance; to those who have given me permission to use their images as illustrations, i.e. the British Library, the British Museum, English Heritage Photographic Library, the National Portrait Gallery, London, Nottingham City Museums and Galleries (Newstead Abbey), and the Harry Ransom Research Center of the University of Texas at Austin; to the editors of *Romanticism* for permission to reproduce some of the material in Chapter Five which was originally published in *Romanticism* 4.1 (1998); to Shirley, Keith and Victoria Dobson, Bill Norbury, Malcolm Smith, Susan and John Willmington and Alan Young for general support, and to Daisy and Muffin for being model companion animals.

King's College London
March 2001

Introduction

'Animals are good to think with'

The whole question of the dog's relation to the spirit of the age, whether it is possible to call a dog Elizabethan, another Augustan, another Victorian, together with the influence upon dogs of the poetry and philosophy of their masters, deserves a fuller discussion than can here be given it.[1]

This book takes Virginia Woolf at her half-serious word, by developing the 'question' posed in *Flush* in terms of the spirit of the Romantic age, and by discussing it in connection not only with dogs but with animals in general at the end of the eighteenth and the beginning of the nineteenth centuries.[2]

Like Woolf's comments, and in conformity with practice in the period which it addresses, this study is primarily concerned with animals as objects in human culture. Claude Lévi-Strauss's provocative observation that 'Animals are good to think with' demonstrates how not only anthropologists but human beings of all kinds tend to conceptualize and quantify animals as a part of the human world, projecting upon them categories and values derived from human society and then serving them back as a critique or reinforcement of the human order.[3] Keith Tester, summarizing Lévi-Strauss, defines a core activity of human civilization as the attempt to say what it is to be properly human: to establish human uniqueness in contradistinction to the otherness of the natural environment.[4] According to this view, culture tries to make sense of, and to control, a world which human beings did not make and to which they have only recently discovered they are not central.

Woolf's concept of 'the spirit of the age' is one which we might now want to qualify, and I return to it in the conclusion to this study: but it is evident that it was in Romantic-period Europe and North America – and especially in Great Britain in this period – that humankind first seriously began to question its own centrality to the world in relation to animals. This is an issue on which we feel ourselves even more deeply uncertain at the opening of the twenty-first century, giving rise to the increasingly active movements in Western societies connected with animal rights and ecological and environmental issues, which are seeking to protect non-human animals from being entirely subordinated to human interests. Already in the Romantic period Samuel Taylor Coleridge's insistence that 'every thing has a life of its own' indicates an awareness of a position which animal rights campaigner Tom Regan expresses more specifically though less elegantly when he claims that animals 'are the

---

[1] Virginia Woolf, *Flush: A Biography* (London: Hogarth Press, 1947) p. 162. The preceding sentence reads: 'Some hold that Byron's dog went mad in sympathy with Byron; others that Nero was driven to desperate melancholy by associating with Mr Carlyle'.

[2] The use of the term 'animal' in this book generally includes birds, fish and insects.

[3] Claude Lévi-Strauss, *Totemism,* trans. by Rodney Needham (Boston: Beacon Press, 1962) p. 89.

[4] See Keith Tester, *Animals and Society: The Humanity of Animal Rights* (London: Routledge, 1991) p. 32.

experiencing subjects of lives that matter to them as individuals, independently of their usefulness to others'.[5]   These modern preoccupations may be seen, however, as yet another way of trying to define and reinforce humankind's position and role in the world in opposition to that of animals. Environmental movements may primarily serve human ends, and Tim Ingold sees 'the Western cult of conservation' as one which 'proclaim[s] that from now on it shall be man who determines the conditions of life for animals (even those still technically wild shall be 'managed'), and who shoulders the responsibility for their survival or extinction.'[6] Even 'kindness' to animals is suspect if, as Marion Scholtmeijer points out,

> humanity in fact serves its own feeling of superiority by taking pity on animals. What if gentleness to nonhuman creatures is the ultimate expression of humankind's pretensions to moral dominance over nature? Animals are not kind to one another, and human beings are, after all, animals themselves. If the law of survival of the fittest applies, then human animals violate nature in countermanding the scheme that has set them up as masters of other animals.[7]

While acknowledging the validity of these late-twentieth-century positions for an age in which humankind has come close to dominating the earth, my study nevertheless identifies itself with its primary and most of its secondary sources by focusing on the way animals are used in human, particularly written, culture. It is only by exploring the nature of our own, current, values and what have been called our 'operative fictions' that we can put ourselves in a position to change them if we wish.[8]

The Romantic era is a rich source of such material because it was then, in the context of a new emphasis on nature, that debate was intensely articulated both about animals' difference from human beings and also about their similarity. In the late-eighteenth century, animals came to be seen as different in that they exist as independent entities from humankind, rather than its mere tools or adjuncts; but they were also perceived as similar, in so far as they have the ability to behave, to feel and perhaps to think like human beings. The industrialization of large areas of the countryside and the movement of population to the towns began to be countered by strong statements by poets in particular about the importance of the natural world, at a time when the

---

[5]   Letter to William Sotheby, 10 September 1802, about the poems of William Bowles: *Collected Letters of Samuel Taylor Coleridge*, ed by Earl Leslie Griggs, 6 vols (Oxford: Oxford University Press, 1956–57) II. 864. Tom Regan, 'Honey Dribbles down your Fur', in *Environmental Ethics: Philosophical and Policy Perspectives* ed by Philip P. Hanson (Burnaby BC: Institute for the Humanities/SFU Publications, 1986) pp. 99–113 (p. 105).

[6]   Tim Ingold, 'Introduction' to *What is an Animal?* ed by Tim Ingold (London: Hyman, 1988) p. 12.

[7]   Marion Scholtmeijer, *Animal Victims in Modern Fiction: From Sanctity to Sacrifice* (University of Toronto Press, 1993) p. 85.

[8]   Steve Baker, *Picturing the Beast: Animals, Identity and Representation* (Manchester: Manchester University Press, 1993) p. 19.

growing accessibility and popularity of verse enabled the pronouncements of poets to be read by ever-widening audiences.[9]

According to Keith Thomas, in his important social-historical work of 1983 (*Man and the Natural World*), this was a development from the situation in early modern England where

> the official concept of the animal was a negative one, helping to define, by contrast, what was supposedly distinctive and admirable about the human species. ... The brute creation provided the most readily-available point of reference for the continuous process of human self-definition. Neither the same as humans, nor totally dissimilar, the animals offered an almost inexhaustible fund of symbolic meaning.[10]

Thomas demonstrates how the change occurred in the three centuries up to 1800 from a view of animals as essentially a part of nature to be subdued for Man's use, to one which both recognized animals' independent standing and acknowledged a strong fellowship between humankind and other species. He traces how the growth of empirical methods in scientific investigation began to demolish the idea of a natural world redolent with human analogy and symbolic meaning, sensitive to Man's behaviour. He illustrates how biological taxonomies such as those of the Comte de Buffon and Carolus Linnaeus delineated anatomical likenesses between the bodies of human beings and those of animals, and how a combination of the perception of physical likeness and a new orientation towards duties of care and the promotion of happiness led to humanitarian movements which addressed the well-being of animals as well as those of slaves and oppressed people. Thomas also claimed that increasing urbanization in the seventeenth and eighteenth centuries reduced close human dependence on animals. This in turn, he believed, had the effect of making animals more respected by town-dwellers, opening up opportunities for essentially bourgeois anti-cruelty movements, promoting an interest in vegetarianism and encouraging the keeping of pets.[11]

---

[9] Francis Jeffrey wrote in 1819: 'There never was an age so prolific of popular poetry as that in which we now live. ... The last ten years have produced, we think, an annual supply of about ten thousand lines of good staple poetry – poetry from the very first hands that we can boast of – that runs quickly to three or four large editions – and is as likely to be permanent as present success can make it.' Review of Thomas Campbell's *Specimens of the British Poets*, in the *Edinburgh Review* 31 (March 1819) in *Jeffrey's Criticism: A Selection*, ed by Peter F. Morgan (Edinburgh: Scottish Academic Press, 1983) pp. 27–8. Henry Maitland wrote in 1817: 'there is at present abroad throughout the world ... a passion for poetry, and more especially for poetry in which the stronger passions of our nature are delineated': *Blackwood's Edinburgh Magazine* 1 (1817) p. 393.

[10] Keith Thomas, *Man and the Natural World: Changing Attitudes in England 1500–1800* (London: Allen Lane, 1983), p. 40.

[11] Hilda Kean disagrees with this: Thomas's analysis, she says, 'fails to recognize the abundance of animals living in cities in the early nineteenth century and their economic, as well as cultural, importance for the inhabitants.' See *Animal Rights: Political and Social Change in Britain since 1800* (London: Reaktion, 1998) p. 30.

Byron and other Romantics

It is with pets, and their death, that this book begins, by deploying in the first
chapter several different readings of one of the Romantic period's most famous
(or notorious) animal poems – Lord Byron's 'Inscription on the Monument of
a Newfoundland Dog' – as a means of presenting material which is relevant as
a background to this study as a whole. This chapter traces the tradition of
'theriophily' (a term drawn by George Boas from the Greek *therios,* a wild
animal) from classical writing up to the Romantic period, in order to explore
the sources on which Byron drew and to provide a philosophical background to
literary approaches to animals in general in late-eighteenth- and early-
nineteenth-century English writing.[12] Theriophily makes an important
contribution to satire, since it consists in demonstrating animals' superiority
over humankind by praising animals' instinctive wisdom and placing human
beings in a sceptical and derogatory light; but an exploration of its history and
development also elucidates wider questions concerned with the theological,
philosophical and political status of animals which were hotly debated in the
Romantic period.

The combination of politics, poetry and Byron's work in the introductory
chapter illustrates two features which are evident throughout this study. The
first of these is the way in which Byron appears as both a typical exemplar of,
and a polarized opposite to, many of the characteristic approaches to animals
in his era. Byron's perception of the consubstantiality – the confraternity or
kinship – between humankind and animals was a feature of his irony and satire
and also caused him to be sympathetic to animals and suspicious of the over-
arching claims of humankind in a way which became increasingly prevalent
through the first half of the nineteenth century. The evolutionary debate about
the scientific relationship between the human and other animal species began
in earnest in the eighteenth century, and it had already established during the
Romantic period that animals and human beings were likely to share a
biological heritage. It was Charles Darwin's contribution to a long-standing
tradition of study to demonstrate, in 1859, not so much the similarity of species
as the mechanism which was the origin of their differences. While, however,
Byron's attitude was one which prefigured Darwinism, it set him at odds with
Utilitarian and other social movements of his time which were primarily
concerned with the 'improvement' of the lower classes, and so urged human
beings to pity and be kind to animals precisely in order to show how *different*
from the 'beasts' they were. In attempting to introduce legislation to ban bull-
baiting in 1802, for example, William Wilberforce described it in Parliament
as 'a practice which degraded human nature to a level with the brutes'.[13]

Byron's deployment of animals in his work also throws into relief attitudes
to nature which are sometimes taken to be typical of English Romantic
literature in general, but which are actually much more characteristic of the

---

[12] See George Boas, *The Happy Beast in French Thought of the Seventeenth Century*
(1909, repr. New York: Octagon Books, 1966) and Leonora Cohen Rosenfield, *From Beast-
Machine to Man-Machine: Animal Soul in French Letters from Descartes to La Mettrie* (1940,
repr. New York: Octagon Books, 1968).

[13] *The Parliamentary History of England, from the Earliest Period to the Year 1803*
(London: Hansard, 1820) XXXVI . 845.

approaches of the first – Wordsworthian – generation than of the 'younger Romantics'. Byron, Shelley and Keats all show their 'belatedness' by being to some degree in flight from the Wordsworthian belief in a firmly-grounded, wild, *green* nature which 'never did betray / The heart that loved her' ('Tintern Abbey', 122–3).[14] In many of his major works Shelley moves away from such a view of nature by taking to the air in free-wheeling panoramas; he fantasizes nature in dreamlike landscapes which are anything but 'natural', and he applies science to theorize and anatomize the natural world. Keats turns his lack of deep experience of rugged natural landscape to defiant advantage by celebrating the garden-like scenery of bowers and 'grots': miniaturizing the picturesque and reifying nature by travelling not literally but literarily through 'realms of gold'.[15] Wordsworth wrote that the 'tone ... of enthusiastic admiration of Nature, and a sensibility to her influences' was one '*assumed* rather than natural' with Byron, and Byron responded with a dismissal of 'this cant about nature' and an impatience with 'this "Babble of green fields" and of bare Nature in general' that was aimed at 'the Lakers', and at Wordsworth in particular.[16] Jonathan Bate remarks how,

> Oddly, in view of their love of all things 'natural', the Lakers and their followers didn't seem to have a particular affection for animals. ... Wordsworth's world at its moments of intensity is strangely silent: the vision enters the boy when the owls don't reply. Nothing could be further from the chatter, screech and rapid movement of Byron's monkey-house.[17]

Chapter One discusses further the place of humour in Byron's deployment of animals and his rejection of the 'Wordsworth physic' about Nature with which, he claimed, Shelley 'dosed [him] even to nausea' in Switzerland in 1816; while the hint from Bate about Wordsworth and animals is developed as part of Chapter Five.[18]

---

[14] *William Wordsworth: Poetical Works,* ed by Ernest de Selincourt and Helen Darbishire, 5 vols (Oxford: Clarendon Press, 1940–49).

[15] See Marjorie Levinson, *Keats's Life of Allegory: The Origins of a Style* (Oxford: Basil Blackwell, 1988, repr. 1990) pp. 11–15.

[16] Wordsworth's letter to Henry Taylor December 26, 1823, in *The Letters of William and Dorothy Wordsworth: The Later Years, part one, 1821–8,* ed by Alan G. Hill, 2nd edn (Oxford: Clarendon Press, 1978) p. 237; Byron's 'Letter to John Murray Esq.' (part of the Bowles/Pope controversy) see *Lord Byron: The Complete Miscellaneous Prose,* ed by Andrew Nicholson (Oxford: Clarendon Press, 1991, hereafter referred to as *LBCMP)* pp. 146 and 136. For references to the 'Lakers and their tadpoles', see *Byron's Letters and Journals,* ed by Leslie A. Marchand, 13 vols (London: John Murray, 1973–94, hereafter referred to as *BLJ)* VI. 10, VIII. 201, and IX. 117.

[17] Jonathan Bate, *The Song of the Earth* (London: Picador, 2000) p. 186. For Byron's monkeys, see for example, *BLJ* VI. 171. The chattering of monkeys is perceived as part of the bewildering and meaningless 'hubbub' of London in *The Prelude* VII. 668.

[18] Thomas Medwin, *Medwin's Conversations of Lord Byron,* ed by Ernest J. Lovell Jr (Princeton: Princeton University Press, 1966) p. 11.

Cultural contexts

A second unifying feature of this study is the way in which each chapter brings
together poetical or self-consciously literary work with other genres of
contemporary writing in order to consider approaches to animals in the period:
so that aspects of animal presentation are compared and contrasted in different
cultural contexts, in poetry and prose, and in imaginative and factual work.
Thus Chapter Two, 'Children's Animals', juxtaposes writing for children at the
end of the eighteenth century with some of Coleridge's poetical and literary
critical work, in the context of contrasts between Lockean and Rousseauian
philosophical and educational theory about children and animals. Chapter
Three, 'Political Animals', considers the 'bullfight stanzas' in Canto I of
Childe Harold in the light of the earliest British parliamentary debates about
cruelty to animals (some of which Byron participated in) and the Burkean
ideas deployed in them. Chapter Four, 'Animals as Food' brings modern
anthropological work on taboos and food, and early-nineteenth-century
vegetarian discourses, to bear on Shelley's idealistic writing and non-animal
eating and Byron's troubled carnivorousness and ruminations on diet.
'Animals and Nature' examines Wordsworth's 'proto-ecological' credentials
by exploring his approach to animals (which he regarded as outside 'Nature')
and by juxtaposing the deployment of animals in his poetry with Romantic-
period concepts about the 'economy of nature' and twentieth-century
ecological ideas. Finally, Chapter Six, 'Evolutionary Animals', shows how
Keats, Byron and others found in pre-Darwinian scientific evolutionary
concepts a rich source of ideas, metaphors and provocative hypotheses about
animals which they could use as the basis for imaginative and creative
exploration in their literary work. Each chapter addresses a different aspect of
animal presentation, and the texts under scrutiny have as far as possible been
permitted to follow their own agenda, rather than being fitted into a
previously-conceived over-arching historical theme or pattern. Several themes
do, however, emerge strongly throughout the work and these are summarized
in the conclusion.

Real animals?

My aim in this study is to show not only *what* was written and said, but
specifically *how* it was written and said: to look in detail at the construction,
rhetoric, allusiveness, figuration, and the imaginative and aesthetic features –
in a word, at the art – of individual texts, as well as the ideological contexts in
which they were set. Animals not only mean very different things to different
people, even within a defined historical period, but they are also used in widely
different technical ways by individual writers and artists: as symbols,
metaphors, fictional protagonists, rallying points, emotional supports, means of
defining the human, insults, the subjects of humour and wit, and objects of
veneration.
    This literary approach is at odds with many other cultural studies of
animals, which have deliberately excluded art and literature from their

discussions. Harriet Ritvo, for example, claims that her own 1987 study of nineteenth-century animals

> present[s] interpretations based primarily on texts produced by people who dealt with real animals – the records of organizations concerned with breeding, veterinary medicine, agriculture, natural history, and the like. ... Canonical art and literature have provided only occasional corroborative examples: the large literature of animal fable and fantasy, which has little connection with real creatures, none at all.[19]

The question here, as Steve Baker points out (p. 14), is what constitutes a 'real' animal? Does the way animals are treated in the scientific disciplines Ritvo names here necessarily make them more 'real' than those featured in 'canonical art and literature'? A historical perspective undermines the notion of a constant 'scientific' approach to the natural world, and highlights the great difference between Romantic-period conceptions of science and those of today. The way animals were characterized in Erasmus Darwin's writing was quite different from the way they appeared in the thinking of his grandson, and that again is distinct from modern scientific depictions. More importantly, we have to recognise that, as Baker says (p.10), culture 'does not allow unmediated access to animals themselves', and for this reason cultural history, cultural anthropology and many other areas of contemporary theory 'are loath to draw a sharp distinction between representation and reality, or between the symbolic and the real' (p. 10). If it is impossible for humans to reach or represent the 'real' animal anyway, is science (with its claim to objectify animals) likely to bring us any closer to them than art (which can at least attempt to treat them as subjects)? One of the central questions in current, twenty-first-century approaches to animals is concerned with the representation of non-human species in political and other arenas. By an analogy with 'minority' human interests, such as those connected with race, gender, disability, and sexual orientation, the concern is that any speaking by the dominant 'majority' on the supposed behalf of the 'minority' risks misrepresenting the minority and undermining its real interests. Andrew McMurry, for example, asks:

> How can you construct nature to 'speak' for itself? Or what amounts to the same thing: how can you allow nature to speak univocally and transparently through you? The answer is that you cannot – unless you are prepared to claim that there is something 'real' beneath the speech,

---

[19] Harriet Ritvo, *The Animal Estate: The English and Other Creatures in the Victorian Age* (Cambridge, Mass.: Harvard University Press, 1987) p. 4. Ritvo's more recent work is much more overt about the relativity with which animals are approached in human culture: 'In a large and complex society, such as that of eighteenth- and nineteenth-century Britain, animals performed many functions and stood (or flew or swam) in relation to many different groups of people. Each of the ways that people imagined, discussed and treated animals inevitably implied some taxonomic structure. And the categorization of animals reflected the rankings of people both figuratively and literally, as analogy and as continuation. That is, depending on the circumstances, people presented themselves as being like animals, or as actually being animals.' Harriet Ritvo, *The Platypus and the Mermaid and other Figments of the Classifying Imagination* (Cambridge, Mass.: Harvard University Press, 1997) p. xii.

untouched and untouchable by the anthropocentric bias, that your own purified and transparent speech merely announces ... In fact, [animals'] inscrutability, their stubborn otherness, may be their greatest protection from the recklessness which so often accompanies our desire to dissect and domesticate a world we have long wished would divulge its 'innermost secrets'. ... While some may take it as a measure of ethical maturity and biocentric magnanimity to make the attempt, it is in fact no great favor to bid non-humans to speak, for even if we did imagine them doing so I suspect we would continue to use them in uncounted ways. The ability to speak has never been a guarantee of confraternity. ... In fact, by defining the subject as one who speaks (as opposed, for example, to one who is read) we are still playing the game of defining non-humans according to some criteria of what we take to constitute the human. ... Why do we suppose that we must ask of nature that it produce a noise, a sign, so that we can demonstrate that it has agency and interests?[20]

A good reason for studying animals as represented in the culture of another period is that the very 'past-ness' of the ideas, the 'shock of the old', and the feeling that the past is indeed 'another country', can give us an almost Brechtian sense of alienation or detachment, which helps us in turn to be aware of and to define the peculiarities of our own *mentalités* or 'operative fictions', which otherwise might be invisible to us.[21] I also suggest there are reasons why the study of selfconsciously-literary and overtly creative writing may actually have an advantage over cultural history, anthropology, sociology and scientific disciplines in representing animals in this sense, specifically *because* it makes no claim to speak of anything more 'real' than human imaginings. First, the intentionally metaphorical nature of the literary work can sometimes throw into relief the less-consciously displayed figurative elements of some of the thinking about animals in the 'non-literary' texts, making the assumptions about animals in the latter writings seem less self-evident and straightforward. When, for instance, the authors of the entry about animals in the 1797 edition of the *Encyclopaedia Britannica* speak of the 'kingdoms' of animals and vegetables, they are, in that one apparently casual word, submitting nature to a system of government which is characteristically human and essentially hierarchical in its outlook.[22] Second, by being open and self-conscious about its metaphorization of animals and the rhetoric with which it manipulates them, literary writing may in fact be more honest about how far human beings can represent animals in any way at all. Imaginative and designedly metaphorical writing may offer a space in which human creativity can experiment with

---

[20] Andrew McMurry, '"In Their Own Language": Sarah Orne Jewett and the Question of Non-human Speaking Subjects', in *Interdisciplinary Studies in Literature and Environment* 6.1 (Winter 1999) pp. 51–63 (pp. 59–62). In Chapter Five I suggest that some of Wordsworth's preoccupations with animals as speaking subjects were operating in the same area.

[21] See Baker, p. 6, referring to Robert Darnton's *The Great Cat Massacre and Other Episodes in French Cultural History* ((London: Penguin, 1985): 'the study of *mentalités* might be broadly defined as "the cultural analysis of popular behaviour and attitudes"'.

[22] *Encyclopaedia Britannica*, 3rd edn, 18 vols (Edinburgh: A. Bell and C. Macfarquhar, 1797) II. 22.

different ideas about animals, without claiming for itself a specious (speciesist?) 'rightness' or 'correctness' about what is being done.

Each writer may be viewed as having their own individual 'operative fiction' about animals; each may be seen to conceive and create animals in a different, non-exclusive – and also plural – way: capable of being differently read by different readers. And in the mean time, the 'very special kind of ritual frame' provided by literary form and discipline, that 'marks off' this kind of writing from other experience (in Mary Douglas's words) is protecting this representational space from being taken for 'reality'.[23] Lévi-Strauss's claim about animals being 'good to think with' emphasizes not only animals' capacity for consolidating, challenging and reforming human ideas and cultures, but also their potential for stimulating human creativity, and the resulting 'goodness' – the aesthetic pleasure – which this can provide. So these writers are (in a way which would make sense to the Romantics and in particular to Shelley) drawing readers to think deeply about animals: attracting or arousing us by means of the striking, skilful, original or beautiful ways they write, and stimulating us to consider whether or not the way in which we now share the earth with other species is the way we want to go on doing so in future.[24]

---

[23] Mary Douglas, *Purity and Danger: An Analysis of the Concepts of Pollution and Taboo* (London: Routledge & Kegan Paul, 1966) p. 162 and throughout. These concepts are discussed further in Chapter Four.

[24] Shelley, 'A Defence of Poetry', in *Romanticism: An Anthology*, ed Duncan Wu (Oxford: Blackwell, 1994, repr. 1999) p. 945: 'the pleasure resulting from the manner in which they [poets] express the influence of society or nature upon their own minds, communicates itself to others, and gathers a sort of reduplication from that community.'

Chapter 1

Animals Dead and Alive: Pets, Politics and Poetry in the Romantic Period

When some proud Son of man returns to Earth,
Unknown to Glory, but upheld by Birth,
The sculptor's art exhausts the pomp of woe,
And storied urns record who rests below;
When all is done upon the Tomb is seen,   5
Not what he was, but what he should have been:
But the poor Dog, in life the firmest friend,
The first to welcome, foremost to defend,
Whose honest heart is still his Masters own,
Who labours, fights, lives, breathes for him alone,   10
Unhonour'd falls, unnotic'd all his worth,
Deny'd in heaven the Soul he held on earth:
While man, vain insect! hopes to be forgiven,
And claims himself a sole exclusive heaven.
Oh man! thou feeble tenant of an hour,   15
Debas'd by slavery, or corrupt by power,
Who knows thee well must quit thee with disgust,
Degraded mass of animated dust!
Thy love is lust, thy friendship all a cheat,
Thy smiles hypocrisy, thy words deceit,   20
By nature vile, ennobled but by name,
Each kindred brute might bid thee blush for shame.
Ye! who behold perchance this simple urn
Pass on, it honours none you wish to mourn:
To mark a friend's remains these stones arise,   25
I never knew but one – and here he lies.
(Lord Byron: 'Inscription on the Monument of a Newfoundland Dog')[1]

Byron's lines, written to commemorate the death of a favourite dog, Boatswain [see Figure 1], who died of rabies in autumn 1808, are a parody of the form they seem to emulate: that of the epitaph. The 'Inscription' subverts the epitaphic function of praising the dead by discoursing instead on the degraded nature of Man, and undermines the traditional consolation of epitaphs by questioning the immortality of the soul. It uses a composition in the form of an epitaph to protest about the role of epitaphs and the way they are composed. Written by a twenty-year-old nobleman who is himself 'Unknown to glory, but upheld by birth', the poem sets itself at odds with the epitaphic convention of family loyalty by having the poet denigrate his own kind.

---

[1]   This version of the text is taken from the monument in the grounds of Newstead Abbey, Nottinghamshire.

While, however, the poem's parodic stance owes much to Popean or Swiftian 'Saeva Indignatio', the 'Inscription' is not wholly consistent in its parody and does not conform throughout to Augustan methods.[2] In contrast to the cultivated distance and universalizing tone of his models, Byron uses the 'I' of his last line to make a point of revealing the poem's personal relevance to the author, and 'real', individual, feelings of loss, loneliness and bitterness seem to break through the misanthropic surface. Equally disconcerting to Byron's original readers was the poem's coupling of a serious and apparently personally embittered tone with the fact that it is about an animal: what the 'Inscription' does *not* do is to mock its own feelings for an apparently inappropriate object, nor – given that its predominant vein is ironical and its subject is an animal – ironize *enough*. The poem has in abundance the antithetical quality which was taken to be the predominant characteristic of Romanticism by Isaiah Berlin in the 1960s; the questioning of its own frame of reference which Jerome McGann identified as a key issue of the movement in the 1980s, and the preoccupation with oppositional polarity which Anne Mellor identified in 1993 as the obsession of 'masculine' Romanticism.[3]

Besides its reworking of Augustan models, the poem adds to its deliberate incongruousness by making use of two specific works from other very different traditions of writing. It unpacks Thomas Gray's 'Elegy Written in a Country Church-Yard' (1751) by means of the Earl of Rochester's 'Satyr against Reason and Mankind' (1674–75), and so brings to bear upon a founding example of eighteenth-century elegiac sensibility the seventeenth-century English culmination of the long classical and Renaissance tradition of theriophily: a philosophical stance which satirizes human pretensions by reminding us of our kinship with animals, and contrasts overweening human folly with animals' instinctive wisdom.

This chapter explores the theriophilic tradition as a way not only of grounding Byron's poem but also of delineating some of the literary and philosophical 'animal' precedents available to other writers in the Romantic period. This approach to the 'Inscription' is one of three readings of the poem which relate to its three forms of publication in Byron's lifetime, each of which markedly varied its reception: first as a monumental inscription; second as a contribution to a friend's miscellany of verses, and third as part of a

---

[2] See Swift's epitaph upon himself: 'Ubi Saeva Indignatio / Ulterius/ Cor Lacerare Nequit' ('[he is gone] where fierce indignation can lacerate his heart no more'). Quoted in *Gulliver's Travels,* ed by Louis A. Landa (London: Methuen, 1965) p. vii.

[3] Isaiah Berlin, in *The Roots of Romanticism,* ed by Henry Hardy (Princeton: Princeton University Press, 1999) p. 18, describes Romanticism as 'in short, unity and multiplicity. ... strength and weakness, individualism and collectivism, purity and corruption, revolution and reaction, peace and war, love of life and love of death.' Jerome J. McGann, *The Romantic Ideology* (Chicago: University of Chicago Press, 1983) claims (p. 73) that 'The difference [between an Elizabethan and Romantic poem] lies in the [Elizabethan] poem's unquestioning acceptance of a stable conceptual frame of reference in which its problems can be taken up and explored. In Romantic poems that frame of reference is precisely what stands at issue.' Anne K. Mellor, in *Romanticism and Gender* (New York: Routledge, 1993) p. 3, shows how 'a binary model is already deeply implicated in "masculine" Romanticism. ... At both the theoretical and the psychological level, the women writers of the Romantic period resisted this model of oppositional polarity ... for one based on sympathy and likeness.'

deliberately-provocative, politically-motivated addendum to one of Byron's best-selling poems. Each of these readings also offers a means of exploring a wider background to Romantic-period animals in general. The first, as noted, serves to introduce a history of thought about animals, around the theme of theriophily and animal epitaphs; the second discusses issues of friendship, relationship and pet-keeping, and the third explores some of the political deployments of animals in the Romantic period, as a prelude to a fuller consideration of this topic in Chapter Three.

Gray and Rochester

Byron's many verse and prose references to Gray's *Elegy* show his close knowledge of and respect for the poem:

> Had Gray written nothing but his Elegy – high as he stands – I am
> not sure that he would not stand higher – – it is the Corner-stone of
> his Glory – without it his Odes would be insufficient for his fame.[4]

Byron's 'Inscription' demonstrates its expectation that the reader will be familiar with the 'Elegy' by allusions to key phrases in Gray's work: Gray's 'storied urn' (line 41) is used unchanged, while his 'pomp of power' (33) becomes Byron's 'pomp of woe', and the 'unhonour'd dead' of Gray's poem (57) are recalled in Byron's by the dog who 'unhonour'd falls'. Byron's punning transformation of Gray's 'animated bust' (41) into the 'animated dust' of line 18 introduces a reference of a type frequent in Byron to the way in which God, having formed Adam out of earth, breathed life or soul *(anima)* into him. The entire concept of the worthlessness of monuments to the proud but undistinguished is drawn by Byron from Gray's contrast between the 'trophies', the 'pealing anthem' which 'swells the note of praise', the soothing 'Flattery' (43), and the 'useful toil, / homely joys, and destiny obscure' (28–29) of the country poor. Byron's nobleman, 'Unknown to glory but upheld by birth', is the obverse of Gray's 'village-Hampden', 'mute inglorious Milton' and 'Cromwell guiltless of his country's blood' (57, 59 and 60) whose achievement of 'glory' was prevented by their humble birth; and Byron's dog as a 'kindred brute' parodies Gray's appeal to a 'kindred spirit' in the reader/friend (96). Finally, the apparently personal tone of the last couplet of the 'Inscription' gains meaning if it is read in the light of the closing Epitaph of the 'Elegy', so that Byron's lines are seen to play only half-ironically upon the implications of Gray's passionate use of the word 'friend': placing Byron's dog, as his sole 'friend', in the same relation to the poet that Gray claims for himself to the 'Youth, to Fortune and to Fame unknown', who 'gain'd from Heaven, 'twas all he wish'd, a friend' (124).

---

4 Letter to John Murray Esq. (part of the Bowles/Pope Controversy, 1821), see *LBCMP*, p. 143. For verse allusions to the 'Elegy' see *Lord Byron: The Complete Poetical Works*, ed by Jerome J. McGann, 7 vols (Oxford: Clarendon Press, 1980–93: hereafter referred to as *LBCPW)* I. 403; II. 333; III. 447; V. 703, and V. 757. In November 1815, in a letter to Leigh Hunt, Byron included Gray in the list of those whose 'addiction to poetry' was, like Byron's own, 'the result of "an uneasy mind in an uneasy body"' *(BLJ* IV. 332).

Although he seems to have expected his readers to pick up his allusions to Gray's 'Elegy', it is unlikely that Byron also assumed them to be familiar with his other major source for the 'Inscription'. *Don Juan* VIII. 252 alludes to Rochester's 'Satyr against Reason and Mankind', and there are other references to 'Wilmot's' verse in an 1806 poem and in Byron's letters and journals, but a taste for Rochester seems to have been very much a minority one in this period.[5] These references place the 'Inscription' within a theriophilic tradition which – despite the work of Boas – has not been much noticed or studied in English culture, although it was common in classical discourse, was revived in Renaissance and seventeenth-century French writing, and came into Restoration English literature from this source.

The 'Inscription' is the most compact and complete of Byron's expressions in this theriophilic tradition which he shares with Rochester.[6] Here, for comparison, is Rochester's opening:

> Were I not (who to my cost already am
> One of those strange, prodigious Creatures, Man)
> A Spirit free, to choose for my own share,
> What Case of Flesh and Blood I pleas'd to weare,
> I'd be a *Dog,* a *Monkey,* or a *Bear,*
> Or anything but that vain *Animal*
> Who is so proud of being rational.[7]

Byron's poem shows this part of its ancestry not only through the contrast it shares with Rochester between noble beast and debased Man (*'Beasts* are in their degree, / As wise at least, and better far than he,' Rochester 115–6), and the way it echoes Rochester's emphasis on the misplaced pride and 'vanity' of humankind (Byron's 'man, vain insect' and Rochester's 'vain animal'), but also through their common emphasis on the uniquely human evil trait of betraying one's own kind.[8] So Byron's 'Thy love is lust, thy friendship all a cheat, / Thy smiles hypocrisy, thy words deceit!' compresses Rochester's lines 135–8:

---

[5]   See *LBCPW*, 'Soliloquy of a Bard in the Country, in an Imitation of Littleton's Soliloquy of a Beauty' 32; and *BLJ* IV. 103 and IX. 20. Samuel Johnson, in his *Lives of the English Poets* (1779–81) described Rochester's works as 'not common', and for this reason 'subjoined his verses'. Johnson ascribed Rochester's fame as a writer in his own time to 'the glare of his general character' and commented that 'because this blaze of reputation is not yet quite extinguished ... his poetry still retains some splendour beyond that which genius has bestowed'. *The Lives of the English Poets,* ed by John Wain (London: Dent, 1975) pp. 109–10.

[6]   Anne Righter and Jonathan Bate discuss Byron and Rochester but not the two poets' link through the theriophilic tradition. See Anne Righter, 'John Wilmot, Earl of Rochester' in *Proceedings of the British Academy,* LIII, 1967 (Oxford: Oxford University Press, 1968), and Jonathan Bate, *The Song of the Earth,* pp. 185–7.

[7]   *John Wilmot, Earl of Rochester, The Complete Works,* ed by Frank H. Ellis (London: Penguin, 1994).

[8]   Byron's theriophily does not here extend to insects, but among the more than thirty references to insects in his verse is the famously sympathetic treatment of the scorpion in ll. 421–34 of *The Giaour* which describes the insect's supposed suicide when surrounded by fire. Spiders usually get a bad press from Byron, however, although the Prisoner of Chillon does manage to befriend them.

But man with Smiles, Embraces, Friendship, Praise,
Most humanly his Fellow's Life betraies,
With voluntary Pains his Works distress,
Not through Necessity but Wantonness.[9]

Byron's allusion to slavery echoes Rochester's reference (176–8) to humans
who tyrannize over animals while having the same status as slaves:

Who swolne with selfish vanity, devise,
False freedomes, holy Cheats, and formal Lyes
Over their fellow *Slaves* to tyrannize,

and the two works share a preoccupation with the immortality of the human
and animal soul which, as I show below, is characteristic of theriophilic
traditions going back more than two thousand years. Byron's bitterness about
the denial to the dog of 'the soul he held on earth', and about the claims for
Man's 'sole exclusive heaven', take their cue from the theological
justifications of Rochester's satirized churchman – the 'formal Band and
Beard' (46) – who takes the poet to task in lines 60–65:

'Blest, glorious *Man!* to whom alone kind Heav'n,
An everlasting *Soul* hath freely giv'n;
Whom his great *Maker* took such care to make,
That from himself he did the *Image* take;
And this fair frame in shining *Reason* drest,
To dignifie his Nature, above *Beast.*'

Both poets sought visual as well as verbal means of expressing their
theriophily, and both used their status as noblemen, with funds free to be put to
unorthodox use, to commission artworks intended to provide a
commemoration of themselves alongside that of the animal. The portrait by an
unknown artist of Rochester crowning a monkey with a laurel wreath [see
Figure 2] was matched by the way in which Byron's ostentatious monument
for Boatswain was constructed (as discussed further below) to accommodate
his own coffin also.

Dead or alive: immortality and theriophily

Byron's use of the *Satyr* and theriophily links him not only with Rochester's
individual seventeenth-century aristocratic and rakish brand of *libertinage* and
scepticism, but also with the wider ongoing paradoxical and sceptical
viewpoint in French and English thought that drew upon classical thinking and
was carried forward particularly by the French *philosophes* into the eighteenth

---

[9] Compare Montaigne's 'As to what concerns fidelity, there is no animal in the world so
treacherous as man': 'Apology for Raymond Sebond' in *Essays of Michel de Montaigne,* trans.
by Charles Cotton, 3 vols (London: G. Bell and Sons Ltd, 1913) II. 163.

century.[10] Byron can, indeed, be seen as one of the last exponents of this particular type of sceptical tradition of human abasement. In Canto XV of *Don Juan* he declares himself a through-going Cynic: 'like Diogenes, / Of whom half my philosophy the progeny is' (XV. 583–4). The Cynics were one of the major groups in classical Greek society, also including the Pythagoreans, who opposed the Aristotelian and Stoic denial of rationality to animals. As Richard Sorabji points out, Diogenes

> rejoiced in the nickname of dog which he was given for his rejection of civilised living, and turned it to a number of witty purposes. Moreover, it was this nickname (*kuôn*) which was transferred to the whole Cynic sect (*kunikoi*). Along with this went Diogenes' insistence that animals were in fact superior to humans.[11]

A key part of the classical debate about the status of animals was the question of whether or not they possessed souls. Animists such as Pythagoras (in the sixth century BC) held that animals and humans shared (and exchanged) spirits of the same kind through metempsychosis or the transmigration of souls, and Pythagoras's condemnation of meat-eating, mediated through literary sources such as Ovid's *Metamorphoses*, led to early vegetarians being known as Pythagoreans. The mechanists took the opposite view and held that *both* humans and animals lacked souls and were no more than machines. These two theories placed animals and human beings on a par, but systems such as Aristotle's – which held that Man shared a nutritive soul with vegetables and a sensitive soul with animals, but alone possessed an intellectual or rational soul – set humankind apart and on a different level from other species.

The first century AD Platonist Plutarch was well-known in the eighteenth century for his treatise advocating vegetarianism, 'On the Eating of Meat', which I discuss in Chapter Four. One of his three other treatises in defence of animals – 'Gryllos', or 'Beasts have Reason' – retells the Homeric story of the witch Circe who turned Odysseus' crew members into pigs. In Plutarch's version, however, the pigs' spokesman, Gryllos or Grunter, refuses to be turned back into a man, claiming that animals are much happier than men, and that Odysseus is a fool not to wish to become a pig himself. Gryllos asserts that animal virtues are better, because they involve no toil, that animal skills are superior because they are untaught, and that beasts are more courageous, more temperate, and wiser than human beings (*Moralia* XII. 493–539).

---

[10] The *Satyr on Mankind* draws heavily on Boileau's *Satire sur l'Homme* (Eighth Satire). Ellis, p. 365, points out that the closing couplet of the *Satyr* – 'yet grant me this at least,/ Man differs more from man than man from beast' – echoes Montaigne's 'There is more distance between man and man than there is between man and beast'. Montaigne in turn echoes Plutarch's Gryllos: 'I scarcely believe that there is such a spread between one animal and another as there is between man and man in the matter of judgement, reasoning and memory' ('Beasts are Rational', in Plutarch's *Moralia* XII, trans. by William C. Helmbold (London: Heinemann, 1957) p. 531.

[11] Richard Sorabji, *Animal Minds and Human Morals: The Origins of the Western Debate* (Ithaca: Cornell University Press, 1993) p. 160.

Pliny's *Natural History* provides another classical instance of this aspect of human self-disparagement – echoed by both Rochester and Byron:

> all other living creatures pass their time worthily among their own species. ... fierce lions do not fight amongst themselves, the serpent's bite attacks not serpents, even the monsters of the sea and the fishes are cruel only against different species; whereas to man, I vow, most of his evils come from his fellow-man.[12]

Pythagoras is credited with an argument which became popular again in the eighteenth century, particularly in literature for children: that cruelty to animals leads to cruelty to one's fellow human beings.[13] The Pythagorean argument was pushed further by Porphyry in the third century AD, who enquired in his *Treatise on Abstinence* whether, if animals had speech, humans would dare to commit what would be perceived as fratricide and cannibalism. This 'indirect' duty towards animals was in fact one of the few benign theories with regard to animals which survived from the classical world into the Christian era in Western Europe. Sorabji's impression (p. 204) is that 'the emphasis of Western Christianity was on one half, the anti-animal half, of a much more wide-ranging and vigorous ancient Greek debate.'[14]

The tone of some Christian attitudes to animals was set by Augustine in a treatise of AD 388, *On the Manichaean and Catholic Ways of Life*, which ascribes to Christ the Stoic theory that animals cannot be brought within the human community of justice and law because they lack reason:

> And first Christ shows your [the Manichaeans'] abstention from killing animals and tearing plants to be the greatest superstition. He judged that we had no community in justice *(societas iuris)* with beasts and trees, and sent the devils into a flock of swine, and withered a tree by his curse, when he had found no fruit in it,

and

> [W]e see and appreciate from their cries that animals die with pain. But man disregards this in a beast, with which, as having no rational soul, he is linked with no community of law *(societas legis).*[15]

---

[12] Pliny, *Natural History,* Book VII, in *Pliny, Natural History with an English Translation in Ten Volumes,* trans. by H. Rackham (London: Heinemann, 1942) II. 507–11.

[13] A Renaissance instance of this belief appears in *Cymbeline* I. v. 23–4 where Cornelius warns the Queen, who claims that she intends to use the poison he has given her on animals: 'Your Highness / Shall from this practice but make hard your heart!' Samuel Johnson wrote of this passage: 'The thought would probably have been more amplified, had our author lived to be shocked with such experiments as have been published in later times, by a race of men that have practised tortures without pity, and related them without shames, and yet are suffered to erect their heads among human beings.' *Johnson on Shakespeare: Essays and Notes Selected and Set Forth with an Introduction* by Sir Walter Raleigh (London: Frowde, 1908) p. 181.

[14] Animals as active participants survive mainly in stories from the folkloric fringes of Christianity: St Jerome is said to have retained the life-long loyalty of a lion after plucking a thorn from its foot; St Francis of Assisi tamed the ferocious wolf of Gubbio; Bede's story of St Cuthbert has the saint's feet warmed and cared for by otters, and representations of Christ's nativity in a stable almost always include animals. See Kean, *Animal Rights,* pp. 9–10.

[15] *On the Manichaean and Catholic Ways of Life,* II. xvii. 58–9, quoted by Sorabji, p. 196.

Augustine mustered this argument to counter the beliefs of the fourth-century Manichaean sect, who did not eat meat and whose holy men refrained even from gathering their own vegetarian food. Thomas Aquinas supported the Porphyrian notion that cruelty to animals could lead to cruelty to human beings, but went further than Augustine in drawing a sharp dividing line between beasts and Man. In the *Summa Theologiae* Aquinas claims that intellectual understanding is the only operation of the soul that is performed without a physical organ, and that animals have understanding only through their physical organs.[16] The irrationality of animals thus makes the difference between beasts and Man an unbridgeable gulf, for in a scheme which is concerned with preparing souls for the next world rather than this one, it is this that gives human beings an immortal and animals merely a mortal soul: 'The human form is the highest, with abilities so transcending physical matter that it possesses an activity and ability which physical matter in no way shares: the power of mind'.[17] As E. S. Turner points out, Aquinas's teaching also helped to lead to the mediaeval custom of putting animals on trial since, although they were deemed to have no souls, animals could be possessed by evil spirits and it was therefore legitimate to curse them 'as satellites of Satan instigated by the powers of Hell.'[18] Christian custom continued to follow the Jewish laws of good husbandry which provided a weekly day of rest for domestic animals, and the presentation of Jesus as the Good Shepherd reinforced this caring attitude towards domestic animals at least, while the concept of Christ as the Lamb of God obviated the need for animal sacrifice in Christian communities. The widespread *Physiologus* tradition also interpreted the characteristics of both real and fantastic animals as symbols of Christian qualities. Theologically, however, the main support for animals' dignity in the mediaeval period came not from the Church but from heretics: the 'Pythagorean' or vegetarian Cathars, who believed in the transmigration of souls, and maintained that all warm-blooded creatures possessed souls of equal dignity.[19]

It was not until the Renaissance that the diversity of the arguments in this area that had been expressed in classical culture was again given an airing. Michel de Montaigne's essay, the *Apology for Raimond Sebond,* first published in 1580, reached a wide public. In his deliberately contentious arguments,

---

[16] Marlowe's Faustus (XIX. 176-9) reflects the Tomasian position when he declares: 'all beasts are happy, / For when they die / Their soulś are soon dissolv'd into elements; / But mine must live still to be plagu'd in hell'. Christopher Marlowe, *Doctor Faustus,* ed by John D. Jump (London: Methuen, 1962) p. 56. John Skelton through Jane Scroupe in the second decade of the fifteenth century had, however, 'craved to God / Nothing else than to keep /Phyllype [Sparrow]'s soul from ... pagan underworld'.

[17] Thomas Aquinas, *Summa Theologiae, A Concise Translation,* ed by Timothy McDermott (London: Eyre & Spottiswoode, 1989) p. 113.

[18] E. S. Turner, *All Heaven in a Rage* (London: Michael Joseph, 1964) p. 26.

[19] Malvolio is anxious to prove his sanity by expressing sound theological opinions on this point in his exchange with 'Sir Topas' in *Twelfth Night* IV.ii.49–54:

Clown: What is the opinion of Pythagoras concerning wildfowl?
Malvolio: That the soul of our grandam might haply inhabit a bird.
Clown: What think'st thou of his opinion?
Malvolio: I think nobly of the soul, and no way approve his opinion.

See also *Doctor Faustus* (XIX. 174–6): 'Ah, Pythagoras' metempsychosis, were that true, / This soul should fly from me and I be chang'd / Unto some brutish beast'.

relying on the genre of paradox which sets out to prove the opposite to commonly-held opinions, Montaigne followed Plutarch in maintaining not merely the goodness but also the superiority of animals:

> The most wretched and frail of all creatures is man, and withal the proudest. He feels and sees himself lodged here in the dirt and filth of the world, nailed and riveted to the worst and deadest part of the universe ... and yet his imagination will be placing himself above the circle of the moon, and bringing heaven under his feet. 'Tis by the vanity of the same imagination that he equals himself to God, attributes to himself divine qualities, withdraws and separates himself from the crowd of other creatures, cuts out the shares of animals his fellows and companions, and distributes to them portions of faculties and force as himself thinks fit ... from what comparison betwixt them and us does he conclude the stupidity he attributes to them? When I play with my cat, who knows whether I do not make her more sport than she makes me? we mutually divert one another with our monkey tricks: if I have my hour to begin or to refuse, she also has hers.[20]

Montaigne also introduced a new sceptical theme based on the opening up of the American continent: the superiority of the noble savage over civilized Man, which was to be further developed and elaborated by Rousseau.

Montaigne's ascription of reason to animals, and Pierre Charron's *De la sagesse* (1601), which dwelt even more than Montaigne on the physical advantages of animals and their moral superiority to humankind, were diametrically opposed by Cartesian dualism which denies reason and even capacity for feeling to all except the thinking human subject. René Descartes' *Discourse on Method* (1637) not only provided the most far-reaching and decisive denial of soul, rationality and feeling in animals to have been extant in the Christian era but also advocated that Man's relationship to animals and to nature in general should be that of the master and possessor. Descartes articulated his argument in terms of the animal as an automaton or machine:

> the circumstance that they do better than we does not prove that they are endowed with mind, for it would thence follow that they possessed greater reason than any of us, and could surpass us in all things; on the contrary, it rather proves that they are destitute of reason, and that it is nature which acts in them according to the disposition of their organs: thus it is seen, that a clock composed only of wheels and weights can number the hours and measure time more exactly than we with all our skill.[21]

The account of Cartesianism given by Pierre Bayle, the late seventeenth-century Protestant scholar and philosopher whose ideas were widely used by Byron in his 'metaphysical' dramas, demonstrates how the theory of the

---

[20] 'Apology for Raimond Sebond', *Essays of Michel de Montaigne,* trans. by Charles Cotton, 3 vols (London: G. Bell and Sons Ltd, 1913) II. 134.

[21] René Descartes, *A Discourse on Method,* trans. by John Veitch (London: J. M. Dent, 1994) p. 43.

animal machine was designed to protect orthodox belief about the immortality of the soul:

> This [the debate about whether beasts have souls] you will say, is of small importance to religion. But you are mistaken, some will answer; for all the proofs of original sin, drawn from sickness and death, which children are subject to, fall to the ground, if you suppose that beasts have sensation, they are subject to pain and death, and yet they never sinned. And therefore you argue wrong, when you say, little children suffer, and die, therefore they are guilty; for you suppose a false principle, which is contradicted by the condition of beasts, viz. That a creature which never sinned, can never suffer. This is nevertheless a most evident principle, which flows necessarily from the ideas which we have of the justice and goodness of God. It is agreeable to immutable order, that order from which we clearly apprehend that God never departs. The souls of beasts confound this order, and overthrow those distinct ideas: it must therefore be granted that the automata of Des Cartes very much favour the principles by which we judge of the infinite Being, and by which we maintain Orthodoxy. [22]

This article, on Rorarius's treatise *That Brute Animals Possess Reason Better Than Man,* demonstrates Bayle's technique of pushing his argument to the point where its apparent orthodoxy cracks open to reveal clear paths to scepticism. Bayle's method and tone is echoed by Lucifer's arguments in Byron's *Cain* (quoted in Chapter Six), as is Bayle's ironic questioning about the place of animals in orthodox Christian belief:

> How do we treat beasts? We make them tear one another in pieces for our own diversion; we kill them for our nourishment, and ransack their bowels during their lives to satisfy our curiosity; and all this we do by virtue of the dominion which God has given us over them. ... Is it not cruelty and injustice to subject an innocent soul to so many miseries? The opinion of Des Cartes frees us from all these difficulties. *(Dictionary, IV.* 902)

These extracts from Bayle's work reflect aspects of the response to Cartesianism which had already been articulated during the seventeenth century, including Fontenelle's famous riposte to the idea of animals as clocks:

> But place a dog-machine and a bitch-machine one beside the other and the result may be a third little machine; whereas two watches will be next to each other all their lives without ever making a third watch.[23]

---

[22] *Mr Bayle's Historical and Critical Dictionary; the second edition, to which is prefixed the Life of the Author revised, corrected and enlarged by Mr Des Maizeaux,* 5 vols (London: Knapton etc, 1734: the English translation used by Byron) IV. 902.

[23] Bernard le Bovier de Fontenelle: *Lettres galantes du chevalier d'Her* (1685) XI, quoted in Rosenfield, p. 126 (my translation).

By the early eighteenth century, educated opinion had hardened against Cartesianism, especially in English circles such as that of Alexander Pope, and Viscount Bolingbroke ridiculed it as an

> Absurd and impertinent vanity! we pronounce our fellow animals to be automatic ... or we graciously bestow upon them an irrational soul, something we know not what, but something that can claim no kindred to the human mind. We refuse to admit them into the same class of intelligence with ourselves ... though it is obvious that the first elements of their knowledge, and of ours, are the same.[24]

Pope was himself well-known as the owner of at least two Great Danes named Bounce, and his poems and letters about his dogs are among the first in English to fully anthropomorphize individual pet animals. Pope's 'Bounce to Fop: An Heroick epistle from a dog at Twickenham to a dog at Court' (1736) again picks up the theme of animal souls:

> And tho' no Doctors, Whig or Tory ones,
> Except the Sect of *Pythagoreans,*
> Have Immortality assign'd
> To any beast, but *Dryden's* Hind:
> Yet Master *Pope,* whom Truth and Sense
> Shall call their Friend some ages hence,
> Tho' now on loftier Themes he sings
> Than to bestow a Word on *Kings,*
> Has sworn by *Sticks* (the Poet's Oath,
> And Dread of Dogs and Poets both)
> Man and his works he'll soon renounce,
> And roar in numbers worthy *Bounce.*
> (82–93).[25]

Pope stops just short of stating his own belief in the immortality of animal souls, but he ascribes at least to the 'poor Indian' the thought that 'admitted to the equal sky / His faithful dog shall bear him company' *(Essay on Man* [1733–74] I. 110–11).[26]

The complex irony of Jonathan Swift's presentation of the Houyhnhnms and the Yahoos in *Gulliver's Travels* may be read either as endorsing or as mocking theriophilic tendencies, but Swift's savage indignation against humankind is certainly a feature shared with theriophily, and a letter from Swift to Pope of September 1725 seems to set the whole of *Gulliver's Travels* in this vein:

---

[24] Henry St John, Viscount Bolingbroke, *Letters or Essays Addressed to Alexander Pope Esq.* (posthumous, 1766) quoted by A. Lytton Sells in *Animal Poetry in French and English Literature and the Greek Tradition* (London: Thames and Hudson, 1957) p. xxvi.

[25] Alexander Pope, *Poetical Works,* ed by Herbert Davis, 1st edn (1966) repr. with new introduction by Pat Rogers (Oxford: Oxford University Press, 1978).

[26] Pope's Roman Catholicism may have influenced him: the gulf between humankind and animals was – and is still – much more strongly stated by the Roman than by the Protestant churches.

I have got Materials Towards a Treatis proving the falsity of that
Definition *animal rationale* [a reasoning animal] and to show it
should be only *rationis capax* [capable of reasoning]. Upon this
great foundation of Misanthropy (though not in Timons manner) The
whole building of my Travels is erected.[27]

It is this harsher, satirical tone which also manifests itself in Julien-Offray de la
Mettrie's *L'Homme-Machine* (1747–48) which maintained that, if animals
were 'machines', so too were humans, since the soul could be seen as a
function of the body in humans as well as in animals. Alongside this
essentially libertine tradition, however, there grew up a gentler, more feeling-
based attitude to animals, which is reflected in Voltaire's criticism of
Cartesianism for its sanction of vivisection, in his entry for 'Beasts' in the
Philosophical Dictionary (1764):

How pitiful, and what poverty of mind, to have said that the animals
are machines deprived of understanding and feeling. ... Judge (in the
same way as you would judge your own) the behaviour of a dog who
has lost his master, who has sought for him everywhere with doleful
cries. ... There are some barbarians who will take this dog, that so
greatly excels man in capacity for friendship, who will nail him to a
table, dissect him alive. ... And what you then discover in him are *all
the same organs of sensation that you have in yourself.*[28]

This humanitarianism also informs Pope's stance, based on the Lockean
conception of a universal great chain of being in which Man and beast are but
a link apart, and allows the poet's imagination to turn itself to presenting the
world from a non-human viewpoint:

Nothing is foreign: parts relate to whole;
One all-extending, all-preserving soul
Made beast in aid of man, and man of beast;
All served, all serving: nothing stands alone:
The chain holds on, and where it ends, unknown.
Has God, thou fool! work'd solely for thy good,
Thy joy, thy pastime, thy attire, thy food?
Who for thy table feeds the wanton fawn,
For him as kindly spread the flowery lawn:
Is it for thee the lark ascends and sings?
Joy tunes his voice, joy elevates his wings.
Is it for thee the linnet pours his throat?
Loves of his own and raptures swell the note. ...
While man exclaims, 'See all things for my use!'

---

[27] Letter to Pope from *The Correspondence of Jonathan Swift*, ed by Harold Williams, 5
vols (Oxford: Clarendon Press, 1962–65) III. 103. Gulliver has in effect become an animal –
either a pet or a kind of insect or vermin – in Brobdingnag, long before he becomes identified
with the Yahoos.

[28] Voltaire, *Dictionnaire Philosophique*, vol. XXXV of *The Complete Works of Voltaire*,
ed Christiane Mervaux (Oxford: Voltaire Foundation, 1994) pp. 411–15; quoted in this
translation by Richard D. Ryder in *Animal Revolution: Changing Attitudes towards Speciesism*
(Oxford: Basil Blackwell, 1989) p. 60.

'See man for mine!' replies a pamper'd goose:
And just as short of reason he must fall,
Who thinks all made for one, not one for all.
(Essay on Man III. 21–34, 45–8).[29]

James Thomson's *The Seasons* (1726–30) reflects this identification with
the viewpoint of animals, and a sensitivity to their suffering which is its
corollary:

> Poor is the triumph o'er the timid hare!
> Scared from the corn, and now to some lone seat
> Retired – the rushy fen, the ragged furze
> Stretched o'er the stony heath, the stubble chapped,
> The thistly lawn, the thick-entangled broom,
> Of the same hue the friendly withered fern ...
> Vain is her best precaution; though she sits
> Concealed with folded ears, unsleeping eyes
> By Nature raised to take the horizon in
> And head couched close betwixt her hairy feet
> In act to spring away.[30]
> *(The Seasons:* 'Autumn' I. 401–14)

Christopher Smart's *Jubilate Agno* (1758–63) explores the religious qualities
of animals, but in this case it is their capacity to worship, rather than the
immortality of their souls:

> For I will consider my Cat Jeoffry.
> For he is the servant of the Living God duly and daily serving him.
> For at the first glance of the glory of God in the East he worships in his
> way.
> For this is done by wreathing his body seven times around with elegant
> quickness.
> For then he leaps up to catch the musk, which is the blessing of God
> upon his prayer.
> *(Jubilate Agno,* Fragment B, 695–9).[31]

A common minute observation of and respect for animals, alongside the two
poets' Hebraic and cabalistic interests, links Smart with a later 'mad poet',
William Blake.[32] Although it seems unlikely that Blake was aware of Smart's

---

[29] Pope also expresses (later in the *Essay on Man* [III. 152–64] and in *Windsor Forest* 124–5) the theriophilic idea that the capacity to betray and attack our own species is a uniquely human one.

[30] *The Seasons: James Thomson,* ed by James Sambrook (Oxford: Clarendon Press, 1981). Byron's descriptions and criticisms of hunting and shooting in *Don Juan* cantos XIV and XVI appear to echo 'Autumn' I. 360–501.

[31] *The Poetical Works of Christopher Smart,* vol. I, *Jubilate Agno,* ed by Karina Williamson (Oxford: Clarendon Press, 1980).

[32] Henry Crabb Robinson claimed in his *Reminiscences* (1869) that, after reading the *Songs of Innocence and of Experience,* Wordsworth commented: 'There is no doubt this poor man was mad, but there is something in the madness of this man which interests me more than

work, since it was unpublished until the twentieth century, the similarities between the two poets' treatment of animals are, as Northrop Frye pointed out, sometimes striking:

> For he counteracts the powers of darkness by his electrical skin
> and glaring eyes.
> For he counteracts the Devil, who is death, by brisking about
> the life
> For in his morning orisons he loves the sun and the sun loves him.
> For he is of the tribe of Tiger.
> For the Cherub Cat is a term of the Angel Tiger.
> (*Jubilate Agno*, Fragment B, 719–23).[33]

Pope's poems about his dogs are an example of the new interest in pet-keeping which may have arisen from the urbanization of the human population and its separation from working animals. So too are the many eighteenth-century elegies for pets, such as John Arbuthnot's 'Epitaph for Signor Fido', inscribed on the back of the Temple of the Worthies at Stowe, Buckinghamshire; Lady Stepney's monument and verses for 'Serpent, a favorite dog' who died in 1750; the verses inscribed by the Earl of Carlisle (Byron's guardian) on the 'monument of a favourite spaniel'; John Gay's 'Elegy on a Lap-Dog'; Oliver Goldsmith's 'Elegy on the Death of a Mad Dog'; Gray's 'Ode on the Death of a Favourite Cat drowned in a Tub of Gold-fishes', and Horace Walpole's epitaph for his little dog Rosette, sent to Lord Nuneham with the expectation it would appeal to his Lordship's 'dogmanity'.[34]

The pet-poem was not, however, new to the eighteenth century: elegies and epitaphs for pets – even locusts, cicadas and grasshoppers kept by children – were common from the fourth century BC. Ovid's *Amores* II. vi (an elegy for Corinna's parrot), Catullus's verses on Lesbia's sparrow, which use the bird as a pretext for interaction between human lovers, and Martial's poem on Publius's little bitch 'Issa', were also influential on English pet-literature. Such material from its earliest classical form often had a satirical intention, and its characteristic tone is light and emotionally detached. Some of it – such as Matthew Prior's 1693 'Epitaph on True, Her Majesty's Dog' – shares Rochester's and Byron's purpose of satirizing humankind in relation to animals:

---

the sanity of Lord Byron and Walter Scott!' Quoted in *William Blake, Songs of Innocence and Experience: A Casebook*, ed by Margaret Bottrall (London: Basingstoke, 1970) p. 37.

[33] 'When Christopher Smart is shut into a madhouse with no company except his cat Jeffrey [sic], the cat leaps into the same apocalyptic limelight as Blake's tiger': Northrop Frye, 'Blake after Two Centuries,' *University of Toronto Quarterly*, xxvii (1957) quoted by Bottrall, p. 167.

[34] Letter to Lord Nuneham, 6 November 1773, quoted by Christopher Hawtree in *The Literary Companion to Dogs* (London: Sinclair-Stevenson, 1993) pp. 130–1. Frederick Howard, Earl of Carlisle, *Poems* (London: J. Ridley, 1773). See further examples of animal elegies and monuments in Nicholas B. Penny, 'Dead Dogs and Englishmen' in *The Connoisseur*, 192. 774 (August 1976) and in Penelope Curtis and others, *Hounds in Leash: The Dog in 18th and 19th Century Sculpture* (Leeds: Henry Moore Institute, 2000).

Ye Murmerers, let *True* evince,
That Men are Beasts and Dogs have Sence.
His Faith and Truth all *White-hall* knows,
He ne're could fawn, or flatter those
Whom he believ'd were *Mary's* Foes. ...
Read this ye Statesmen now in Favour,
And mend your own, by *True's* Behaviour.
(15–22).[35]

Most eighteenth-century pet epitaphs similarly have an approach marked by wit, a mock-heroic stance and an absence of affect; and it is relatively late in the century before a new sensibility enters such compositions, perhaps taking its cue from the incident of the traveller grieving over his dead ass in Laurence Sterne's *A Sentimental Journey* (1768):

And this, said he, putting the remains of a crust into his wallet – and this should have been thy portion, said he, hadst thou been alive to have shared it with me. I thought by the accent, it had been an apostrophe to his child; but 'twas to his ass, and to the very ass we had seen dead in the road. ... The mourner was sitting upon a stone bench at the door, with the ass's pannel and its bridle on one side, which he took up from time to time – then laid them down – look'd at them and shook his head. He then took his crust of bread out of his wallet again, as if to eat it; held it some time in his hand – then laid it upon the bit of his ass's bridle – looked wistfully at the little arrangement he had made – and gave a sigh. ... Shame on the world! said I to myself – Did we love one another, as this poor soul but loved his ass – 'twould be something.– [36]

In Cowper's epitaphs on his pet hares Tiney and Puss (1783) the old question of the status of animal souls again becomes an issue.[37] Cowper's

[35] *The Literary Works of Matthew Prior,* ed by H. Bunker Wright and Monroe K. Spears (Oxford: Clarendon Press, 1959). This tradition of using animals to satirize the Court sometimes extends the satire to the monarch and the animals themselves, as in Frances Coventry's (anonymous) *History of Pompey the Little: or, The Life and Adventures of a Lapdog* (London, 1751) p. 1: '[I]f we descend to later times, neither there shall we want examples of great mens devoting themselves to dogs. King Charles the second, of pious and immortal memory, came always to his council-board accompanied with a favourite spaniel; who propagated his breed, and scattered his image through the land, almost as extensively as his royal master. His successor, King James, of pious and immortal memory likewise, was distinguished for the same attachment to these four-footed worthies; and 'tis reported of him, that being once in a dangerous storm at sea, and obliged to quit the ship for his life, he roar'd aloud with a most vehement voice, as his principal concern, "to save the dogs and colonel Churchill"'.

[36] Laurence Sterne, *A Sentimental Journey through France and Italy by Mr Yorick* ed by Gardner J. Stout Jr, (Berkeley: University of California Press, 1967) pp. 138–41.

[37] Tiney died in March 1783, aged eight years and five months. Puss died in March 1786, aged nearly twelve, but his epitaph seems to have been written when he was nine years old. See *The Poems of William Cowper,* ed by H. John Baird and Charles Ryskamp, 3 vols, (Oxford: Clarendon Press, 1995) II. 312–13.

poem on Tiney remains carefully orthodox, and the fact that it promises no immortality to his pets adds to its touchingness:

> But now, beneath this walnut-shade
>   He finds his last, long home,
> And waits in snug concealment laid,
>   'Till gentler Puss shall come.
>
> He, still more aged, feels the shocks
>   From which no care can save,
> And, part'ner once of Tiney's Box,
>   Must soon partake his grave.
> ('Epitaph on a Hare', 37–44)

The Latin epitaph for Puss is also non-Christian in its tone, and the effect of the implications for the poet of Puss's death is distanced and mediated by the classical language:

> Tamen mortuus est –
>   Et moriar ego
>
> [Nevertheless, he died –
>   And I too shall die]
> (*Epitaphum Alterum* 12–13, my translation).[38]

Wordsworth, however, in his first 'Essay on Epitaphs' (1810) makes it clear that, for him, it is this very lack of immortal expectations for animals which makes it inappropriate to provide them with epitaphs. Although his argument does not specifically mention Christian views, it clearly draws upon them:

> The dog or horse perishes in the field, or in the stall, by the side of his companions, and is incapable of anticipating the sorrow with which his surrounding associates shall bemoan his death, or pine for his loss; he cannot pre-conceive this regret, he can form no thought of it; and therefore cannot possibly have a desire to leave such regret or remembrance behind him. Add to the principle of love which exists in the inferior animals, the faculty of reason which exists in Man alone; will the conjunction of these account for the desire? Doubtless it is a necessary consequence of this conjunction; yet I do not think as a direct result, but only to be come at by an intermediate thought, viz. that of an intimation or assurance within us, that some part of our nature is imperishable.[39]

---

[38] Cowper was also a master of the mock-heroic animal epitaph: examples include 'Θn the Death of Mrs Throckmorton's Bullfinch', and 'To the Immortal Memory of the Halybutt on which I dined this day.' The latter has a classical nuance – according to James Serpell, *In the Company of Animals: A Study of Human-Animal Relationships* (Oxford: Basil Blackwell, 1986) p. 37: 'At one time pet turbot were all the rage in Rome. The daughter of Drusus adorned one with gold rings, while the orator Hortensius actually wept when his favourite flatfish expired.' Anna Seward's 'An Old Cat's Dying Soliloquy' is in the same vein.

[39] *The Prose Works of William Wordsworth,* ed by W. J. B. Owen and Jane Worthington Smyser, 3 vols (Oxford: Clarendon Press, 1974) II. 50.

Similarly, the occasion of the death of a favourite spaniel, Music, in 1805 provided Wordsworth with an opportunity to meditate upon and carefully demarcate the difference between the commemoration of animals and of human beings:

> Lie here, without a record of thy worth,
> Beneath a covering of the common earth!
> It is not from unwillingness to praise,
> Or want of love, that here no Stone we raise;
> More thou deserv'st; but *this* man gives to man
> Brother to brother, *this* is all we can.
> ('Tribute to the Memory of a Favourite Dog', 1–6)

Diverse as they are, the animal elegies and epitaphs I have mentioned, including Byron's, Cowper's and Wordsworth's, all move quite swiftly away from commemoration of the dead animal to the concerns of the living poet, and use the animal's death as a pretext for exploring differences – rather than similarities – between human beings and animals. Byron's 'Inscription' generalizes only briefly and in clichéd terms about 'the poor dog', and culminates in an abrupt change of subject in the last couplet to focus intensely instead on the personal problems of the poet. Cowper does individualize his pet, but he is also quite specific about why Tiney was kept at all:

> I kept him for his humour' sake,
>   For he would oft beguile
> My heart of thoughts that made it ache,
>   And force me to a smile.
> (33–6)

Wordsworth also describes Music's character, but mainly in order to demonstrate how 'human' and therefore *un*like other dogs she was, and thus to explain why it is acceptable for men and women to make an exception by mourning her:

> Yea, for thy fellow brutes in thee we saw
> The soul of love, love's intellectual law.
> Hence, if we wept it was not done in shame,
> Our tears from passion and from reason came,
> And therefore shalt thou be an honoured name!
> ('Tribute', 32–6)

Likewise, True's demise provides the opportunity for the poet satirically to point a moral for politicians. Although, as the eighteenth century progresses, animal epitaphs increase in their wish and ability to show and invoke feeling about the animal's death, they continue to demonstrate a feature which is in general characteristic of approaches to animals in pre-Darwinian imaginative literature: when they look at other species it is essentially with the purpose of commenting on our own, and in order to address issues of human culture by comparing the human with the animate non-human.

Pets, friends and other animals

As mentioned above, the first 'publication' of Byron's 'Inscription' was on a large and elaborate stone monument erected above the spacious underground chamber where Boatswain was buried [see Figure 3]. The monument was commissioned by Byron for the grounds of Newstead Abbey, his ancestral estate in Nottinghamshire, and the monument and its inscription are still very visible there today. Byron left instructions in his will of 1811 that when the time came he was to be buried in this tomb himself, alongside his dog and his old servant, Joe Murray, and there is a slab inside the tomb of an appropriate size to carry human coffins.[40] The context of the monument thus enacts various themes which are also prominent in the poem: its theriophily; its calculated defiance of Christian traditions about burial and memorial; its dissent from Christian teaching about the immortality of the soul, and the consubstantiality between human beings and other animals.[41] Against the dark sandstone background of the Gothic ruins of the Abbey church, the predominantly pale neo-classical monument, erected by a young aristocrat (as yet 'ennobled but by name' himself), cocks a very public snook at orthodox religion and tradition through the construction of a piece of satirical sculpture which mocks aristocratic monuments and 'names' not himself but (in a separate prose inscription) the dog. The verses explain what is going on. The reader/observer is, however, aware of the irony of the fact that only an aristocrat with an ancestral estate and considerable wealth would have had the ability to erect and maintain such a monument – and to commandeer his servant into participating in this piece of performance art.[42]

The verse on the monument is preceded by a prose epitaph:

Near this spot
Are deposited the Remains of one
Who possessed Beauty without Vanity,
Strength without Insolence
Courage without Ferocity,
And all the Virtues of Man without his Vices.
This praise, which would be unmeaning Flattery
If inscribed over human Ashes

---

[40] The tomb was opened and examined on 25 September 1987. See *Lord Byron's Newstead* by Rosalys Coope (no publishers' details, 1988), Appendix 1: 'Byron's vault at Newstead'. 'The chamber itself is barrel-vaulted and measures a fraction over 10 ft square. ... Projecting into the centre of the chamber out of the rear wall is a stone slab, 6 ft long and 3 ft wide ... clearly designed to receive a coffin far larger than one made for a dog.'

[41] For Byron's views on immortality at this time see his letter to Francis Hodgson, 3 September 1811: 'I will have nothing to do with your immortality; we are miserable enough in this life, without the absurdity of speculating upon another' *(BLJ* II. 88).

[42] See *The Works of Lord Byron: Poetry,* ed by Ernest Hartley Coleridge, 7 vols (London: John Murray, 1898, hereafter referred to as *WLBP)* I. 280. 'In the will which the poet executed in 1811, he desired to be buried in the vault with his dog, and Joe Murray was to have the honour of making one of the party. When the poet was on his travels, to whom a gentleman, to whom Murray showed the tomb, said, "Well, old boy, you will take your place here some twenty years hence." "I don't know that, sir," replied Joe; "if I was sure his lordship would come here I should like it well enough, but I should not like to lie alone with the dog."'

Is but a just tribute to the Memory of
BOATSWAIN, a Dog,
Who was born at Newfoundland, May 1803,
And died at Newstead, Novr. 18th 1808.

In its original form, this was not written by Byron, and the writing of the prose epitaph constituted, in fact, a different kind of performance, illustrating another theme which is prominent in the poem: that of friendship. The poem states that the poet has no friends except the dog, but the contribution of the prose epitaph, written in its original form by Byron's Cambridge friend John Cam Hobhouse, commemorates precisely Hobhouse's support for Byron in mourning the dog and enacts what was in any terms the exceptionally loyal friendship Hobhouse was to provide for Byron throughout his life and after his death.[43] Hobhouse related the incident leading to its composition in a draft letter intended for the *New Monthly Magazine* in 1830, when planning to defend Byron's memory against charges of 'bitterness of spirit' which had been detected in the dog's memorial:

> Another fact is that the epitaph was written not by Lord Byron, but by myself – Lord Byron had shown me his verses on the death of Boatswain – which your readers will recollect conclude thus –
> 'To mark a friend's remains these stones arise
> I never knew but one – and here he lies.'
> On reading them, I suggested an alteration which I told my friend substituted worse grammar but better sense – it was to insert 'I' for 'he' in the last line – on which Lord Byron burst out laughing & exclaimed – 'why you are not jealous of the dog are you?' My rejoinder was – 'Perhaps I am a little – but I can praise him in prose and match your misanthropy' – accordingly I wrote the epitaph and Lord Byron directed it to be engraved together with his own verses on the monument now in Newstead gardens.[44]

---

[43] Hobhouse's original version of the epitaph seems to have been even longer and more elaborate. Ralph Lloyd-Jones records, in 'Boatswain is dead!' in *The Newstead Abbey Byron Society Newsletter* XII (Winter, 1997) pp. 3–9, his discovery of the following version of it in Hobhouse's handwriting in the John Murray archives: 'In this consecrated ground lie the remains of a favourite, Who was so singular in his Fortune as to have many Friends, Being of a nature truly heroic, and not less capable of making than of fearing an enemy. Although his Temper was kind, engaging, and in a peculiar degree domestic, his courage was undaunted, and brought him off Victorious, in Fifty Pitched Battles. He died in the fifth year of his age, but lived long enough for his glory, and to give such undoubted Proofs of Valour, Generosity, and of every engaging Quality, that he was as lamented in his death, as he was loved in his life. This praise, which would be unmeaning Flattery if inscribed over human ashes, is but a just tribute to the memory of Boatswain, a Dog, who was born at Newfoundland, and died at Newstead, On Wednesday the 11th of November, 1808.' (Note the different death date given).

[44] Quoted in Doris Langley Moore, *The Late Lord Byron, Posthumous Dramas* (London: John Murray, 1961) pp. 366–7. Thomas Moore describes the 'Inscription' as 'misanthropic' in his *Letters and Journals of Lord Byron with Notices of his Life,* 2 vols (London: John Murray, 1830–31) I. 154, but goes on to contextualize Byron's dog-love by describing Pope's eulogy for Bounce and the ancient Greek dog's grave at Salamis, and by citing Hume's account of how Rousseau's dog 'acquired the ascendant' over him, Burns's elegy on the sheep, Mailie, Cowper's fondness for his spaniel Beau and Sir Walter Scott's celebration of his dog Maida.

The same process is demonstrated by the poem's second publication, which was in *Imitations and Translations, From the Ancient and Modern Classics, Together with Original Poems never before Published,* by J. C. Hobhouse, 1809.[45] In this scenario, Hobhouse authorizes a poem by his friend Lord Byron, who claims that his only real friend is the dog. Byron's (or Hobhouse's) awareness of the oddity of the performance enacted by this publication is indicated by the fact that the last line was changed to read 'I knew but one unchang'd – and here he lies', although this actually seems an even more pointed assertion of Byron's preference for the dog.[46] Hobhouse's motives for involving himself so heavily in Byron's mourning for Boatswain may indeed be an expression of his attempts to master his own jealousy by intervening in and taking possession of Byron's expression of grief: Hobhouse gave evidence later of so urgent a need to be in control of some aspects of Byron's life that Doris Langley Moore suggests *(Late Lord Byron,* pp. 12–56) that he even proposed and organized the burning of Byron's Memoirs out of jealousy that they had been given by Byron to Thomas Moore and not to himself.

The second publication of the 'Inscription' can thus be compared to the enactment of the 'erotic triangle' of 'male homosocial desire' defined by René Girard and Eve Kosofsky Sedgwick, where within the male-centred novelistic tradition of European high culture, a pattern is uncovered whereby (usually) two men are rivals for (usually) one woman, and where 'the bond that links the two rivals is as intense and potent as the bond that links either of the rivals to the beloved'.[47] In the narrative of the Boatswain memorial, the dog takes the place of a lover, as the two men rival each other in expressions of admiration and grief, and the animal's death thereby provides literary pretexts for exploring the men's friendship, mutual patronage and opportunities for the fulfilment of 'homosocial desire'.

The introduction of an element of desire or eroticism into the supposed relationship of Byron and his animals is not a twentieth-century invention. When Shelley described Byron's Ravenna household in a letter of 1821 to Thomas Love Peacock he mythologized a setting which had evidently disturbed him by its elements of 'polymorphous perversity'. Referring to the residence as 'this Circean palace', he told Peacock that

> Lord Byron's establishment consists, besides servants, of ten horses, eight enormous dogs, three monkeys, five cats, an eagle, a crow, and a falcon; and all these, except the horses, walk about the house,

---

[45] *Imitations and Translations* was published 'about the time Hobhouse and Byron left Falmouth in June 1809 to enact Childe Harold's first journey', according to Donald H. Reiman in his edition of the work: *John Cam Hobhouse, 'Imitations and Translations and The Wonders of a Week at Bath'* (New York: Garland, 1977) p. vi. The monument and inscription were in existence by April 1809, when Charles Skinner Matthews left Newstead after commenting on the inscription (see *BLJ* VII. 225).

[46] In Byron's final publication of the poem (1814) the original version of this line was restored.

[47] Eve Kosofsky Sedgwick, *Between Men: English Literature and Male Homosocial Desire* (New York: Columbia University Press, 1985) p. 21.

which every now and then resounds with their unarbitrated quarrels, as if they were the masters of it.[48]

As he wonders 'who all these animals were before they were changed into these shapes', Shelley characterizes Byron as a Circean enchantress who has turned her former lovers into beasts – perhaps envisaging that the working-class mistresses and (possibly) male lovers, with whom he had last seen Byron associating in Venice, had actually been metamorphosed here into the animals to which Byron himself had admiringly compared them.[49]

The involvement with animals entered 'Byronism' – that mythic version of the poet's persona and biography – at an early stage in his lifetime, in the form of a little manuscript book of ink-and-water-colour drawings and hand-written verses in the style of 'Old Mother Hubbard': 'The Wonderful History of Lord Byron & His Dog' (1807), composed by Byron's friend and neighbour in Southwell, Elizabeth Pigot.[50] Pigot's pictures show the nineteen-year-old Byron sitting at a table writing; steaming in a slipper-bath; returning from eating oysters [see Figure 4], and riding in a coach, while Boatswain ('Bosen') performs various tricks; and this friendly approach still crops up regularly in anthologies and magazine articles.[51] Its first unfriendly appearance was in a lampoon on Byron as part of the Cambridge 'fast set' published by Hewson Clarke in *The Satirist* in 1808, and Clarke's verses already imply a strong element of sexual ambiguity in the relationship between Byron and his animal:

Sad Bruin, no longer in woods thou art dancing,
　With all the enjoyments that Love can afford;
No longer thy consorts around thee are prancing,
　Far other thy fate – thou art slave to a Lord!

---

[48] *The Letters of Percy Bysshe Shelley,* ed by F. L. Jones, 2 vols (Oxford: Oxford University Press, 1964) II. 331.

[49] Shelley wrote of Byron's Venetian life: 'He allows fathers & mothers to bargain with him for their daughters & though this is common enough in Italy, yet for an Englishman to encourage such sickening vice is a melancholy thing. He associates with wretches who seem almost to have lost the gait and physiognomy of man, & who do not scruple to avow practices which are not only not named but I believe seldom conceived in England. He says he disapproves, but he endures.' *(Letters* II. 58). Byron described his Venetian mistress Margarita Cogni ('La Fornarina' or the baker's wife) as a 'gentle tigress', 'energetic as a Pythoness' and 'a magnificent animal' *(BLJ* VI. 192–7).

[50] The second edition of John Harris's *The Comic Adventures of Mother Hubbard and her Dog,* published in 1805, had, like Pigot's parody, illustrations coloured by hand, and was one of the first fiction books published with colour for young children. Four of the illustrations from Pigot's book are reproduced in Annette Peach, 'Portraits of Byron' reprinted from *The Walpole Society,* LXII (2000) figs 6–9.

[51] For contemporary Byroniana on animals, see *Medwin's Conversations of Lord Byron* pp. 3 and 10; *Lady Blessington's Conversations of Lord Byron,* ed by Ernest J. Lovell Jr (Princeton: Princeton University Press, 1969) p. 40, and Thomas Moore, *Letters and Journals,* I. 92. For modern Byroniana on this theme see G. L. Pendred, 'Mad, Bad and Dangerous Dogs,' in *Country Life,* 182. 10 (March 10, 1988) pp. 156–7; Christine Smith, 'Lord Byron: A Dog's Best Friend' in *Kennel Gazette,* August 1991, and Margaret Brown, '"The Firmest Friend"' in *Tails of the Famous* by Elizabeth Edwards and Margaret Brown (Bourne End: Kensal, 1987) pp. 13–29.

How oft when fatigued, on my sofa repining,
  Thy tricks and thy pranks rob of anger my breast;
Have power to arouse me, to keep me from dosing [sic],
  Or what's the same thing, they can lull me to rest.

But when with the ardours of Love I am burning,
  I feel for thy torments, I feel for thy care;
And weep for thy bondage, so truly discerning,
  What's felt by a Lord *may be felt by* a BEAR![52]

Clarke's lines refer, of course, to another of Byron's pieces of performance art involving animals: to his keeping a tame bear at the university (the authorities forbade undergraduates to keep dogs) and his oft-repeated joke that it should 'sit for a fellowship':

> There is a youth ycleped Hewson Clarke ... [who] for no reason that I can discover, except a personal quarrel with a bear, kept by me at Cambridge to sit for a fellowship, and whom the jealousy of his Trinity cotemporaries [sic] prevented from success, has been abusing me, and what is worse, the defenceless innocent mentioned, in the Satirist for one year and some months. (Postscript to *English Bards and Scotch Reviewers* [1809])[53]

Squibs such as Clarke's contain a range of innuendo which it is now difficult to pick out: in earlier centuries, however, bears had been associated with 'phlegmatic' characteristics ('inertness, coldness, beastliness and unfeelingness') and with the element of water: so that there may perhaps be a reference to Byron's habit – then considered as odd as his ostentatious pet-keeping – of swimming in the Cam.[54] Keith Thomas points out (p. 39) how in the seventeenth century it was considered 'bestial' to go swimming: 'for, apart from being in many Puritan eyes a dangerous form of semi-suicide, it was essentially a non-human method of progression. As a Cambridge divine observed in 1600: men walked, birds flew, only fish swam'. George Canning sounded the same theme with a political edge in 'The Progress of Man' in *The Anti-Jacobin* in 1798:

> First – to each living thing, whate'er its kind,
> Some lot, some port, some station is assign'd.
> The Feather'd Race with pinions skim the *air* –
> Not so the Mackerel, and still less the Bear: ...

---

[52] *The Satirist* II (June 1808) (London: Samuel Tipper) p. 368.

[53] One of the *Oxford English Dictionary's* definitions of 'fellowship' (3.c) is 'sexual intercourse'. Anne Barton points out, in 'Lord Byron and Trinity: A Bicentenary Portrait' in *Trinity Review,* 1988 (Cambridge: Trinity College, 1988) pp. 3–6 (p. 3), that this 'was a joke with a cutting edge. Although Byron's tutor Jones had successfully pressed, some years before, for fellowship elections to be conducted openly rather than in private, they were still susceptible to charges of favouritism and abuse.'

[54] See Clive Bush, *The Dream of Reason: American Consciousness and Cultural Achievement from Independence to the Civil War* (London: Edward Arnold, 1977) p. 310.

But each, contented with his humble sphere,
Moves unambitious through the circling year;
Nor e'er forgets the fortunes of his race,
Nor pines to quit, or strives to change, his place.
Ah! who has seen the mailéd Lobster rise,
Clap his broad wings, and soaring claim the skies?
When did the Owl, descending from her bow'r,
Crop, 'midst the fleecy flocks, the tender flow'r?
Or the young Heifer plunge with pliant limb
In the soft wave, and fish-like strive to swim?
(32–49)

A note to the last five lines points out that 'Every animal [is] contented with the lot which it has drawn in life. A fine contrast to Man – who is always discontented.'[55]

Clarke's primary emphasis seems to be on Byron's and the bear's 'companionship' (perhaps with a play on the bear's College 'fellowship') and the fact that they are alike – in their common subjugation to 'the ardours of love', perhaps in the ungainliness of their gait (with a reference to Byron's lameness), and in their tendency to 'dose' on sofas.[56] But innuendoes of bestiality are also distinctly present in the bear's power to both 'arouse' and 'lull ... to rest' the young lord. Certainly, these lines, and the accompanying prose which suggested that Byron's talents best fitted him to keeping a bear-garden, and that 'he has been seen to hug [his 'favourite'] with all the warmth of *fraternal* affection!' were provocative enough to lead Byron (who had been carrying on a 'pure but passionate' affair with the Cambridge choir-boy, John Edleston) into attempting to challenge Clarke to a duel.

Boatswain and the Cambridge bear were only the most notorious of Byron's 'companion animals', however, and throughout his life he was an habitual and self-reflective pet-keeper and an acute observer and recorder of animal activity of all kinds. Byron's pet-keeping habits were bourgeois in their origins (he and his mother lived modestly in an apartment in Aberdeen when he was a child, but they seem always to have owned dogs): later, however, they became distinctly aristocratic in their scale. Besides the Italian menagerie detailed by Shelley, Byron promoted many unlikely animals into pets, including four Greek tortoises (and the hen he 'hired' to hatch their eggs), a hedgehog, and three Italian geese, originally intended for a Michaelmas dinner.[57] The

---

[55] *The Anti-Jacobin*, January 1798 (London: J. Wright, 1798). There is an allusion here to the dietary laws of *Leviticus*, which are further discussed in Chapter Four.

[56] The bears' ungainliness had a specific cause. A speaker in the parliamentary debate on bull-baiting in 1802 described how they were taught to dance: 'by putting a hot iron under the hind feet of the bear, and holding him in an upright position, and playing a tune on the fiddle; the pain made the animal move quickly about and change his position as quickly as possible ... the burning pain became so associated with the music, that whenever a fiddle was played, the same impression was conveyed to his brain, and produced similar gesticulations, and correspondent motions in his feet.' (*Parliamentary History* XXXVI. 842).

[57] *BLJ* II. 48, II. 106. The geese became the subject of puzzled correspondence between Byron's Italian banker Barry and Byron's executors: 'He had kept for a long time three common geese,' Barry wrote, 'for which, he told me, he had a sort of affection, and

establishment of the menagerie mimicked collections such as the one of lions, leopards and other ferocious beasts which, from the twelfth century until 1834, were kept by the monarchs of England at the Tower of London, or the ostriches maintained by the Duke of Cumberland in Windsor Great Park. Keith Thomas points out (p. 277) how the royal menagerie 'symbolized its owner's triumph over the natural world; some medieval rulers even demonstrated their valour by fighting against their captive beasts. Later the zoo became a symbol of conquest as well as of wealth and status'.[58] Byron's large dogs and Cambridge bear seem to have expressed the power and defiance of authority that the young lord would have liked to wield but was constrained from doing in reality, while the menagerie in Italy enhanced his status as a grand and eccentric English 'milord'.[59]

Byron's monkeys and birds were, however, also opportunities for tenderness and concern:

> The crow is lame of a leg – wonder how it happened – some fool trod upon his toe, I suppose. The falcon pretty brisk – the cats large and noisey – the monkeys I have not looked to since the cold weather, as they suffer by being brought up. ('Ravenna Journal', 6 January 1821, *BLJ* VIII. 15)

and took the place (along with his servants) of the family from which he felt he had been ousted by the enforced separation from Lady Byron:

> How is all your rabbit-warren of a family? I gave you an account of mine by last letter. The Child Allegra is well – but the Monkey has got a cough – and the tame Crow has lately suffered from the head ache. – – Fletcher has been bled for a Stitch – & looks flourishing again. – Pray write. (Letter to Augusta Leigh, 18 November, 1820, *BLJ* VII. 2)

Although as I have indicated a good deal of biographical material has been amassed on the subject of Byron and animals, there has been surprisingly little critical commentary on this aspect of his work. Jonathan Bate's discussion of animals and Byron's humour in 'The Song of the Earth' (mentioned in my Introduction) is, it appears, one of only two critical notices of this subject; the other being G. Wilson Knight's comments about Byron's 'in-feeling into animal life and energy' and his perception of 'the poet's deep sympathy for animals, often small ones (as in Pope), which is ... one with his penetration to

---

particularly desired that I should take care of them. [...] I will send them to England, if you please'. They were still alive and still in Italy in 1827. See Doris Langley Moore, *The Late Lord Byron*, p. 61.

[58] In France, 'the animals kept by the king in his menagerie at Versailles were removed by the revolutionary government in 1792 to the Jardin des Plantes in Paris, so that the people could see the animals in public and free of charge' (Kean, p. 23).

[59] According to John Berger, 'Why look at animals?' in *About Looking* (London: Writers and Readers, 1980) pp. 1–26 (p. 12), the pet-owner's relation to the pet is inherently corrupt: 'The pet *completes* him, offering responses to aspects of his character which would otherwise remain unconfirmed'.

the central energies, the springs of action, in beast or man'.[60] This may be because Byron's own references to animals are themselves mainly in the form of short, often ironic, asides, which serve as a constantly sardonic commentary on human affairs and indicate a persistently pro-animal point of view but which are never elaborated or developed into any coherent philosophy. 'Dogs have such intellectual noses!' we are told in the shipwreck scene of Canto II of *Don Juan* (461), just before this particular dog – Juan's 'small old spaniel' – falls prey to soon-to-be-cannibalistic human hunger. 'Hounds, when the huntsman tumbles, are at fault' (VIII. 128) the narrator comments quizzically, in one of the many hunting metaphors in the 'battle' cantos (eight and nine) of the poem; and in the same place bears and wolves are theriophilically compared favourably with the ferocious 'Cossaques' from whom Juan rescues the child Leila (VIII. 730–6). Here also even 'the poor Jackals' (IX. 214–16) are 'less foul' than 'that mercenary pack all, / Power's base purveyors, who for pickings prowl'. Canto IX (146–52) also provides a novel gloss on the Biblical Fall, from the animals' point of view, which was later more widely developed by Byron in *Cain:*

> We have
> Souls to save, since Eve's slip and Adam's fall,
>   Which tumbled all mankind into the grave,
> Besides, fish, beasts, and birds. 'The Sparrow's fall
>   Is special providence,' though how *it* gave
> Offence, we know not; probably it perched
>   Upon the tree which Eve so fondly searched.

Byron claimed to have ceased participating in field sports himself in 1808 *(BLJ* I. 181), and to have given up shooting at live targets after wounding an eagle in Greece in 1810 *(BLJ* III. 253), and he makes clear his disdain for hunting, shooting and fishing in the Norman Abbey cantos of *Don Juan.* There he execrates not only the 'deadly shots, Septembrizers keen' who go searching for 'the poor partridge through his stubble screen' (XVI. 683), but also the anglers with their ambiguous 'solitary vice' (XIII. 845), whose patron, Isaac Walton, 'The quaint, old, cruel cockscomb, in his gullet / Should have a hook and a small trout to pull it', as well as the foxhunters (XIV. 278), of whose activities Juan

> After a long chase o'er hills, dales, bushes,
> And what not, though he rode beyond all price,
> Ask'd next day, 'If men ever hunted twice?'[61]

---

[60] G. Wilson Knight, 'The Two Eternities,' in *Modern Critical Views: George Gordon, Lord Byron,* ed by Harold Bloom (New York: Chelsea House Publishers, 1986) p. 41.

[61] There appears to be an echo here of Thomson's condemnation of angling with a worm in 'Spring' 388–93 from *The Seasons,* although Thomson approves of fly-fishing. Byron's note to the passage on fishing *(LBCPW* V. 759) is equally scathing about Walton: 'This sentimental savage, whom it is a mode to quote (amongst the novelists) to show their sympathy for innocent sports and old songs, teaches how to sew up frogs, and break their legs by way of experiment, in addition to the art of angling, the cruellest, the coldest, and the stupidest of pretended sports. They may talk of the beauties of nature, but the angler merely

Elsewhere in this book I discuss several of the more developed animal themes in Byron's work, but one aspect of this subject which demands separate comment is the extent to which Byron's disability and his empathy with animals are connected. What animal bodies and deformed human bodies have in common is being denied the 'normal' human privilege of being made in the image of God: Byron's Calvinistic upbringing made him well aware of his exclusion in this respect, and the sense of identification with animals which permeates his work provides a means of challenging not just the Christian orthodoxy that only God-like human bodies are available for salvation but also Romantic claims about the uniqueness of human sublimity. In Chapter Six I explore how Byron's chosen nickname for himself, 'le diable boiteux', draws on the eponymous hero of Alain-René Lesage's novel – half-dwarf and half-goat – to construct a satirical and defiant alternative to authority and orthodoxy through a disabled body, and whenever in his work animal, or mixed animal-human, bodies are compared with 'normal' human ones, the comparison consistently results in the advantage being given to those with animal features.[62] In his 1811 journal Byron considers only half-facetiously the benefits, in the next life, of having *'four* legs by way of compensation' for the 'bodily inferiority' of his lameness *(BLJ* II. 47), and in *Mazeppa* this wish is sublimated in the wild, dream-like sequence where boy and horse combine to become one centaur- or Pegasus-like creature:

Away, away, my steed and I,
  Upon the pinions of the wind,
  All human dwellings left behind;
We sped like meteors through the sky.
(422–6)[63]

Byron made common cause with animals for all sorts of iconoclastic projects, including casting himself in his role as creative artist as the diabolically grotesque 'ape' or imitator of God.[64] His deployment of the Grotesque (for example, in *Don Juan* and *The Deformed Transformed)* demonstrates his understanding of this genre as one which mixes human and animal characteristics – and beautiful and deformed human features – in playful but provoking and sometimes deeply disturbing forms. Victor Hugo's

---

thinks of his dish of fish. ... No angler can be a good man.' This seems to be a dig at Hobhouse, who was particularly fond of fishing.

[62]   Alain-René Lesage, *Le Diable Boiteux: texte de la deuxième édition avec les variantes de l'édition originale et du remaniement de 1726,* ed by Roger Laufer (Paris: Mouton, 1970).

[63]   Byron's journal entry in Malta: 'Besides in another existence I expect to have *two* if not *four* legs by way of compensation,' strikingly recalls Tom Paine's sceptical observation about bodily immortality in 'The Age of Reason' *(Thomas Paine: Representative Selections,* ed by Harry Hayden Clark (New York: American Book Co, 1944) p. 319), 'Besides, as a matter of choice, as well as of hope, I had rather have a better body and a more convenient form than the present.' Paine critically compares Man's physical abilities with those of various animals.

[64]   See Marina Warner, *No Go the Bogeyman* (London: Chatto & Windus, 1998) p. 247:'[grotesque's] capacity to provoke unease has excited reproof throughout its history. ... its very detachment from logic and biology can take a disturbing turn. ... [Vitruvius and Vasari] were sensitive to something diabolical about the arbitrary playfulness of the artist'. See also Bate, *Song of the Earth,* pp. 186–7.

identification (1827) of Grotesque as the distinguishing feature of Romantic art, and of *Beauty and the Beast* as its defining narrative, illustrates the way in which the vitality of the genre was being rediscovered in the early nineteenth century.[65] In our own era, Mikhail Bakhtin's elucidation of the carnivalesque as antipathetic to the bourgeois and the classical, and fundamentally at odds with the finished and completed human body, has ensured the Grotesque's continuing relevance to aspects of modern and postmodern culture concerned with representations of the body.[66] Byron's sense of his 'kinship' with the 'brutes' as one which was body-based and consubstantial undoubtedly gained added piquancy and depth from this aspect of his experience.

Wordsworth and Byron had widely differing stances on the beneficent influence of nature upon humanity. Many critics have perceived Byron's attempts in *Childe Harold* III to imitate Wordsworthian expressions of the inspirational and consolatory and powers of nature as inherently unstable and unconvincing, and Byron later summarily dismissed the 'Wordsworth physic' on this subject with which he said Shelley dosed him in Switzerland in 1816. Ernest J. Lovell agrees with Wordsworth in perceiving in Byron's work an 'inability to formulate for himself an intellectual system satisfactorily relating God, man and nature', or to 'find any deep or abiding satisfaction in a life close to nature'.[67] Lovell's analysis (p. 23) of the three ways Byron approached nature at different periods of his life ('variously picturesque, realistic, or comic') helps to define why Byron appears much less convinced than some of his contemporaries about the spiritual power of inanimate nature. An artist who approaches nature as picturesque engages with the natural world through a deliberately-imposed human frame and is particularly conscious of those aspects of nature which link it to the human world. Realistic representations downplay the numinous and spiritual aspects of nature and may emphasize those features which are inimical to Man; while comic approaches privilege nature's animate over its inanimate aspects. Indeed the whole tenor of Byron's argument in his 'Letter to John Murray Esq' in the Bowles/Pope controversy is that nature without animation is barren for poetic purposes. For Byron the main source of animation is provided by human artworks, such as St Peter's, the Coliseum, the Pantheon, the Venus de Medici, the Dying Gladiator and the ruins of Greece, which are, he claims,

> as *poetical* as Mont Blanc or Mount Aetna – perhaps more so – as they are direct manifestations of mind – & *presuppose* poetry in their

---

[65] 'In modern thought, ... the grotesque has an immense role. It is everywhere within it; on the one hand it creates the deformed and the horrible; on the other the comic and the humorous. It is this ... which gives Satan his horns, his goat's feet, his bat's wings ... In our view the grotesque is the richest source that nature can open up for art. ... Antiquity would never have created *Beauty and the Beast.' Préface de 'Cromwell' de Victor Hugo,* ed by Edmond Wahl (Oxford: Clarendon Press, 1932) pp. 14–19 (my translation).

[66] Mikhail Mikhailovich Bakhtin, *Rabelais and his World,* trans. by Helene Iswolsky (Bloomington: Indiana University Press, 1984). Bakhtin's work has been applied to Byron's major poem by Philip W. Martin: 'Reading *Don Juan* with Bakhtin' in *Don Juan* ed by Nigel Wood (Buckingham: Open University Press, 1993) pp. 90–121.

[67] Ernest J. Lovell Jr, *Byron: the Record of a Quest: Studies in a Poet's Concept and Treatment of Nature* (Austin: the University of Texas Press, 1949) pp. 22 and 55.

very conception – – and have moreover as being such a something of
actual life which cannot belong to any part of inanimate nature. ...
take away Rome – and leave the Tyber and seven Hills ... [and] let
Mr Bowles – or Mr Wordsworth or Mr Southey – or any of the other
'Naturals' make a poem upon them – and then see which is most
poetical – their production – or the commonest Guide-book. ('Letter
to John Murray Esq', 1821, *LBCMP* pp. 134–5)

While human beings and their works provide for Byron the most powerful
element of animation in natural scenes, the perception of other forms of life
(both animal ones and those of a personified 'Nature') is essential to him to
give meaning and coherence to even the most sublime and Wordsworthian
landscape:

> To sit on rocks, to muse o'er flood and fell,
> To slowly trace the forest's shady scene,
> Where things that own not man's dominion dwell,
> And mortal foot hath ne'er or rarely been;
> To climb the trackless mountain all unseen,
> With the wild flock that never needs a fold;
> Alone o'er steeps and foaming falls to lean,
> This is not solitude; 'tis but to hold
> Converse with Nature's charms, and view her stores unroll'd.
> (*Childe Harold* II. 217–25)

The rights of brutes?

This second reading of the 'Inscription' for the Newfoundland Dog monument
has considered animals – primarily in connection with Byron – as pets, friends,
companions, family relations, supposedly as lovers, and as consubstantial
partners in a satirical, comic and realistic approach to life. The third
publication of the 'Inscription' was in the contentious setting of the second
edition of the hugely-popular *Corsair,* early in 1814, which included reprints
of both the 'Inscription' and of the notorious 1812 'Lines to a Lady Weeping'
(previously published anonymously) and thus identified the latter as being by
Byron. The 'Lines' contained an attack on the Prince Regent and led to Byron
being widely criticized in the newspapers. Indeed, the publication together of
the 'Lines' and the religiously-unorthodox 'Inscription' can be taken as a
watershed in the history of Byron's popularity in British high society,
signalling as it did the start of the perception by part of the 'Great World' that
he was 'not one of us'. It was in this context that the theriophily of the
'Inscription' and Clarke's 1808 *Satirist* lampoon were brought together in
verses by 'Unus Multorum' in the *Morning Post* in February 1814 which
hailed Byron as (amongst other things) 'Friend of the dog, companion of the
bear'.[68]

---

[68] For 'Attacks upon Byron in the Newspapers for February and March 1814', see *The Works of Lord Byron: Letters and Journals,* ed by Rowland E. Prothero, 6 vols (London: John Murray, 1903, hereafter referred to as *WLBLJ)* II, appendix VII, pp. 463–92. These include 'To Lord Byron' by Unus Multorum *(Morning Post,* 16 February 1814) p. 485.

My third reading of the 'Inscription' thus uses Byron's poem as a starting-point for a consideration of the use of animals in political contexts in the Romantic period. Becoming the 'companion of a bear' had already been associated as long ago as the 1790s with liberal causes, through a well-known poem by Robert Southey (who by 1814 had moved far away from his youthful radicalism and had become Poet Laureate the year before). Southey's 'The Dancing Bear; Recommended to the Advocates for the Slave-trade' linked sympathy for a dancing bear with the continual failure of Parliament to introduce anti-slavery legislation:

> Alas, poor Bruin! How he foots the pole,
> And waddles round it with unwieldy steps,
> Swaying from side to side! ...
>   Bruin-Bear
> Now could I sonnetize thy piteous plight,
> And prove how much my sympathetic heart
> Even for the miseries of a beast can feel,
> In fourteen lines of sensibility.
> But we are told all things are made for man;
> And I'll be sworn there's not a fellow here
> Who would not swear 'twere hanging blasphemy
> To doubt that truth. Therefore as thou wert born,
> Bruin! for man, and man makes nothing of thee
> In any other way, – most logically
> It follows, that thou must be born to dance.
> (4–6, 16–27)[69]

The poem moves on to a description of the bear's debased condition which parodies arguments in favour of human slavery – 'being born / Inferior to thy leader, unto him / Rightly belongs dominion' (38–40); ''Tis wholesome for thy morals to be brought / From savage climes into a civilized state' (43–44); 'thy welfare is thy owner's interest' (50) – and thence to a lament that, in Parliament, 'For seven long years this precious Syllogism / Hath baffled justice and humanity!' (54–5).

Southey's poem takes us into the area also highlighted in lines 15–18 of the 'Inscription', where animals are associated with underprivileged groups in *human* society, and therefore bound up with movements for these groups' better treatment or liberation.[70] There is a long history of the use of animals in 'political' literary contexts as symbols of oppressed religious groups or of

---

[69] *Poems of Robert Southey,* ed by Maurice H. Fitzgerald, (London: Henry Frowde, Oxford University Press, 1909). Southey was later the author of the famous children's tale, 'The Story of the Three Bears', which was originally recounted to his children and eventually published in *The Doctor* in 1837. His references to animals are consistently kindly: in the first book of *Thalaba the Destroyer* the only human being to be spared by the Icy Wind of Death is a man who has saved a dying camel, and his 'On the Death of a Favourite, Old Spaniel' makes a famously tender-hearted contribution to the tradition of elegies for pets.

[70] See Keith Thomas p. 44: 'Slaves were often given names of the kind normally reserved for dogs and horses. One eighteenth-century London goldsmith advertised 'silver padlocks for blacks or dogs'; and English advertisements for runaway negroes show that they often had collars round their necks.'

legitimate innocence, including Una's 'milk white lamb' in Book One of Edmund Spenser's *The Faerie Queene;* the animals in Andrew Marvell's 'Nymph complaining for the Death of her Faun'; John Dryden's *The Hind and the Panther,* and Pope's *Windsor Forest.* In the late eighteenth century, however, the association of animals with oppressed human groups moved out of the purely symbolic realm and became much more direct. Preachers such as Humphry Primatt and John Wesley developed the religious dimension of the analogy in their sermons. Primatt's 'The Duty of Mercy and the Sin of Cruelty to Brute Animals' (1776) linked cruelty to animals directly with slavery by making difference in appearance no bar to similarity of feeling: so that kindness was due to an animal because it was 'no less sensible of pain than a man', and to slaves since, 'as there is neither merit nor demerit in complexion, the white man, notwithstanding the barbarity of custom and prejudice can have no right, by virtue of his colour, to enslave and tyrannise over a black man.'[71] Wesley (see Kean, pp. 19–21) believed that animals had immortal souls, and he inveighed against both sporting and domestic cruelty, including the 'outrage and abuse' which was Man's return for the 'life and faithful service' rendered to him by 'the generous horse' and 'the faithful dog'. In *A Sentimental Journey* Sterne (who was also a clergyman) similarly linked the starling which 'can't get out' and Yorick's meditations upon human slavery (pp. 197–201).

In the secular sphere, the issue of animal cruelty became associated with questions of rights and citizenship. The liberal and republican ideology of the time distinguished between (on the one hand) arms-bearing, property-owning male citizens, with a physical stake in the state and therefore the right to participate in its decision-making, and (on the other hand) groups – such as women, colonized races, slaves, children and working men without the vote – who were characterized as irresponsible, potentially volatile and closer to nature.[72] At the end of the eighteenth century, attention began to be directed to the extension of rights – or at least liberty – to the latter, subordinated groups. Since animals could be seen to be metonymically or synecdochically linked to these oppressed human groups, they were drawn into the debate, and the continuum of better treatment and rights was also, to some extent, applied to them.[73]

---

[71] Quoted in Kean, *Animal Rights,* pp. 18–19.

[72] See Jean-Jacques Rousseau, *Emile; or, Treatise on Education,* trans. by William H. Payne (London: Edward Arnold, 1901) book V, 'The Education of Women', pp. 259–308; Caroline Franklin, *Byron's Heroines* (Oxford: Clarendon Press, 1992) p. x, and Margaret Canovan, 'Rousseau's two concepts of citizenship', in *Women in Western Political Philosophy: Kant to Nietzsche,* ed by Ellen Kennedy and Susan Mendus (Brighton: Wheatsheaf, 1987) pp. 78–105.

[73] The republican John Oswald reflected the link between political liberty and anti-cruelty measures in the Preface to *The Cry of Nature: an Appeal to Mercy and Justice on behalf of the Persecuted Animals* (London: J. Johnson, 1791, p. ii) : 'when he [the author] considers the natural bias of the human heart to the side of mercy, and observes, on all hands, the barbarous governments of Europe giving way to a better system of things, he is inclined to hope that the day is beginning to approach when the growing sentiment of peace and goodwill towards men will also embrace, in a wide circle of benevolence, the lower order of life.' A modern view of the same idea is expressed by Peter Singer and W. E. H. Lecky, who suggest that the circle of moral concern successively expands from the egoistic individual to the family group,

This integrated approach is demonstrated by Jeremy Bentham's well-known comment of 1789 that

> The day has been, I grieve to say in many places it is not yet past, in which the greater part of the species, under the denomination of slaves, have been treated by the law exactly on the same footing, as, in England for example, the inferior races of animals are still. The day *may* come, when the rest of the animal creation may acquire those rights which never could have been witholden from them but by the hand of tyranny. The French have already discovered that the blackness of the skin is no reason why a human being should be abandoned without redress to the caprice of a tormentor. It may come one day to be recognized, that the number of the legs, the villosity of the skin, or the termination of the *os sacrum*, are reasons equally insufficient for abandoning a sensitive being to the same fate. What else is it that should trace the insuperable line? Is it the faculty of reason, or, perhaps, the faculty of discourse? But a full-grown horse or dog, is beyond comparison a more rational, as well as a more conversible animal, than an infant of a day, or a week, or even a month, old. But suppose the case were otherwise, what would it avail? the question is not, Can they *reason?* nor, Can they *talk?* but, Can they *suffer?* [74]

Conversely, the potentially politically de-stabilizing force of such views is demonstrated by satirical writing such as that of Thomas Taylor ('the Platonist'), whose *A Vindication of the Rights of Brutes* (1792) ridicules the argument that all human beings are equal by proposing that equality and civil rights should be extended to animals (and indeed vegetables and minerals) as well:

> We may therefore reasonably hope, that this amazing rage for liberty will continually increase; that mankind will shortly abolish all government as an intolerable yoke; and that they will as universally join in vindicating the rights of brutes, as in asserting the prerogatives of man. [75]

---

community, nation, humanity and ultimately, animals. See Singer, *The Expanding Circle: Ethics and Sociobiology* (Oxford, Clarendon Press, 1981) p. 121: 'The idea of equal consideration for animals strikes many as bizarre, but perhaps no more bizarre than the idea of equal consideration for blacks seemed three hundred years ago.'

[74] Jeremy Bentham, *The Collected Works of Jeremy Bentham: An Introduction to the Principles of Morals and Legislation,* ed by J. H. Burns and H. L. A. Hart (University of London and Athlone Press, 1970) p. 238. These sentences are a note to the main argument about the need for a utilitarian moral and legal code. Part of Bentham's argument was anticipated by Thomas Hobbes: see *Leviathan* ed by Michael Oakeshott (Oxford: Basil Blackwell, 1957) p. 16, 'It is not prudence that distinguishes man from beast. There be beasts that at a year old observe more, and pursue that which is for their good more prudently, than a child can do at ten.'

[75] Thomas Taylor, *A Vindication of the Rights of Brutes,* ed by Louise Schutz Boas (Gainesville, Florida: Scholars' Facsimiles & Reprints, 1966) p. vii.

Taylor goes on to propose (pp. 77–8) that, since women and girls have been so emboldened by the idea of equal rights with men, they will no longer be afraid of elephants, and that the mating of women and elephants will produce offspring with the advantages of both reason and superhuman strength.

In 1762, anticipating Romantic-period vegetarian arguments, Oliver Goldsmith commented of animal-lovers that 'they pity and they eat the objects of their compassion', and many of Southey's poems about animals participate in this debate by demonstrating arguments of both sides of the question.[76] Southey seems not only imaginatively to enter the animal's viewpoint and engage seriously with the ethics of animal treatment, but also to turn back and ridicule such seriousness. His poems from the 1790s such as 'To a Goose', 'To a Spider', 'The Pig, a colloquial poem', 'Sonnet – the Bee', and 'Ode to a Pig, while his nose was boring' all address the animal familiarly, and seem to identify sympathetically with its feelings:

> Spider! thou need'st not run in fear about
>   To shun my curious eyes;
> I won't humanely crush thy bowels out
>   Lest thou should'st eat the flies;
> Nor will I roast thee with a damn'd delight
>   Thy strange instinctive fortitude to see,
>   For there is one who might
>   One day roast me. (1–8)[77]

Animals such as the College Cat are seen to be natural allies for the young anti-authoritarian student at Oxford:

> For three whole days I heard an old Fur Gown
> Beprais'd, that made a Duke a Chancellor:
> Trust me, though I can sing most pleasantly
> Upon thy well-streak'd coat, to that said Fur
> I was not guilty of a single rhyme!
> 'Twas an old turncoat Fur, that would sit easy
> And wrap round any man, so it were tied
> With a blue riband. ...
>   Swell thy tail
> And stretch thy claws, most Democratic beast,
> I like thine independence! Treat thee well,
> Thou art as playful as young Innocence;
> But if we play the Governor, and break
> The social compact, God has given thee claws,
> And thou hast sense to use them.
> ('To a College Cat, written soon after the installation at Oxford,' 1793, 18–35)

---

[76] Oliver Goldsmith, *The Citizen of the World* (London: Everyman Library, 1934) p. 38.

[77] Southey appears to echo a poem by William Oldys (1696-1761): 'On a Fly drinking out of his Cup', and Blake in 'The Fly' *(Songs of Innocence and of Experience,* 1795) ll. 1–8: *The Complete Poems of William Blake*, ed by Alicia Ostriker (Harmondsworth: Penguin, 1977):

Little Fly,	Am not I
Thy summer's play	A fly like thee?
My thoughtless hand	Or art not thou
Has brush'd away.	A man like me?

The identification of animals' plight with that of slaves and other oppressed human groups – usually satirically expressed – is a constant motif:

The social pig resigns his natural rights
    When first with man he covenants to live;
He barters them for safer stye delights,
    For grains and wash, which man alone can give. ...

And when, at last, the closing hour of life
    Arrives (for Pigs must die as well as Man),
When in your throat you feel the long sharp knife,
    And the blood trickles to the pudding-pan;

And when, at last, the death-wound yawning wide,
    Fainter and fainter grows the expiring cry,
Is there no grateful joy, no loyal pride,
    To think that for your master's good you die?
('Ode to a Pig while his Nose was Being Bored', 17–44)

Such satire works in both directions, however, and although the main point is to show how slaves are treated like animals by their masters, the equation of animals with slaves comes uncomfortably close to suggesting that, if slaves should not be treated like this, neither should animals. An earlier model for this motif may be found in Anna Barbauld's 'The Mouse's Petition to Doctor Priestley' (from Barbauld's *Poems*, 1773):

If e'er thy breast with freedom glo'ed,
And spurned a tyrant's chain,
Let not thy strong oppressive force
A free-born mouse detain. ...

The well-taught philosophic mind
To all compassion gives;
Casts round the world an equal eye,
And feels for all that lives. (9–29) [78]

Southey appears deliberately to lead himself into, and to relish, such dilemmas of anthropomorphism, resulting in a witty but odd variability of tone between satiric humour and sentimental seriousness. In 'The Pig, a Colloquial Poem', for example, the argument is divided between Jacob, whose nose is 'Turn'd up in scornful curve at yonder pig' (l. 2), and the narrator, who elaborates on the animal's 'pig-perfection', and points out that it is

a democratic beast,
[Who] knows that his unmerciful drivers seek
Their profit, and not his. He hath not learnt
That pigs were made for man, – born to be brawn'd
And baconized; that he must please to give
Just what his gracious masters please to take. (10–15)

[78] *The Works of Anna Laetitia Barbauld,* ed by Lucy Aikin, 2 vols (London: Longman, 1825).

Democracy was, however, a concept still to be ridiculed at this period, even by radical students, and just as Jacob seems to be convinced by the narrator's argument, it is the narrator who scents the breeze: 'O'er yon blossom'd field / Of beans it came, and thoughts of bacon rise' (59–60). Similarly, a *Morning Post* poem of 1799, 'Elegy upon Eggs and Bacon', humorously relishes the poet's moral helplessness:

> And I have din'd! again have made my meal –
>> Yes, and I found the Eggs and Bacon sweet;
> Alas! that men who eat should finely feel –
>> Alas, that men who finely feel should eat!
>
> Cease, Sensibility, torment no more!
>> Spare me my bosom's lov'd, yet tyrant Queen!
> Why tell me what the Bacon was of yore,
>> And wherefore portray what the Eggs had been?
>
> Perhaps, poor PORKER! when the butcher came,
>> Thoughts of delight were ripening in his breast;
> First, for some fair he felt the tender flame –
>> She was not coy, and he had sure been blest.
>
> And ye, poach'd Eggs! to life ye soon had burst,
>> With sudden strength and consciousness endued;
> How carefully the hen your youth had nurst,
>> How proudly cackl'd o'er her beauteous brood. ...
>
> But wherefore should I muse on thoughts like these?
>> Why wake the wounds of feeling thus unwise? –
> Nay, nay, ye Eggs and Bacon, be at peace,
>> Nor in my conscience, nor my stomach, rise.

The fellow-feeling between disenfranchised human being and oppressed animal gains much more of a political edge when the poet is not writing from a position of affluent ease and privilege, but feels himself to be a member of a human social underclass, as Burns does:

> I'm truly sorry man's dominion
> Hath broken nature's social union,
> An' justifies that ill opinion
>> Which makes thee startle
> At me, thy poor, earth-born companion
>> An' fellow-mortal! ...
> Still, thou art blest, compared wi' me!
> The present only toucheth thee:
> But och! I backward cast my e'e,
>> On prospects drear!
> An' forward, tho' I canna see,
>> I guess an' fear!
> ('To a Mouse, on Turning Her up in Her Nest with the Plough', 7–48).[79]

---

[79] *The Poetical Works of Burns,* ed by Raymond Bentman (Boston: Houghton Mifflin, 1974).

A stage further on from this is when the poet sympathizes or identifies not only with the oppression, but also with the recalcitrance or defensive energy of the tormented animal, as in Wordsworth's *Peter Bell* or John Clare's badger, cornered by men and dogs:

> He runs along and bites at all he meets
> They shout and hollo down the noisey streets
> He turns about to face the loud uproar
> And drives the rebels to their very doors
> The frequent stone is hurled where ere they go
> When badgers fight and every ones a foe
> The dogs are clapt and urged to join the fray
> The badger turns and drives them all away
> Though scarcly half as big dimute and small
> He fights with dogs for hours and beats them all
> The heavy mastiff savage in the fray
> Lies down and licks his feet and turns away
> The bull dog knows his match and waxes cold
> The badger grins and never leaves his hold
> He drives the crowd and follows at their heels
> And bites them through the drunkard swears and reels.
> ('Badger', 27–42) [80]

If in his own time Clare was valued it was as a 'peasant poet' who was seen as being able to rise above the supposed peasant viewpoint of taking his natural surroundings for granted. In the late twentieth century he is enjoyed more as a poet who resisted the accepted literary presentations of the natural world, and maintained an ability to see it with a startling individuality. This includes his approach to fellow-creatures like the snipe, living and often suffering, like himself, as a recluse on the borders of human society:

> Lover of swamps
> And quagmire overgrown
> With hassock-tufts of sedge, where fear encamps
> Around thy home alone
>
> The trembling grass
> Quakes from the human foot,
> Nor bears the weight of man to let him pass
> Where thou, alone and mute,
>
> Sittest at rest
> In safety, near the clump
> Of huge flag-forest that thy haunts invest
> Or some old sallow stump.
> ('To the Snipe', 1–12)

Clare's portrayals of animals and especially birds (he wrote over forty bird poems) show them as having their own, non-human, preoccupations and

---

[80] *John Clare,* ed by Eric Robinson and David Powell (Oxford: Oxford University Press, 1984).

concerns, and as making their own perilous attempts (shared with the working-class poet) to evade or defy oppression by human kind, in the form of enclosures of common land, insecurity of housing, shooting and hunting.

Blake's 'The Tyger' is poised between celebrating the energy and dreading the destructiveness of an animal that does not need to fear anything from Man:

> Tyger, Tyger, burning bright
> In the forests of the night,
> What immortal hand or eye
> Could frame thy fearful symmetry?
>
> In what distant deeps or skies
> Burnt the fire of thine eyes?
> On what wings dare he aspire?
> What the hand dare seize the fire?
>
> And what shoulder, & what art,
> Could twist the sinews of thy heart?
> And when thy heart began to beat,
> What dread hand? and what dread feet?
> ('The Tyger' 1–12)

Blake's poem benefits from a reading which takes into account the political use of animals in contemporary prose as symbols for disenfranchised people – in their angry and destructive character as well as their passive, oppressed one. The British newspaper reports of the French Revolution and its aftermath frequently deployed animals in this way, and Blake's various drafts of 'The Tyger', which show alternately ferocious and benign representations of the animal, can be associated with the more and less bloody phases of the Terror of 1792–93, as Blake may have read them. *The Times* of 7 January 1792 claimed, for example, that the French people were now 'loose from all restraints, and, in many instances, more vicious than wolves or tigers', while on 26 July 1793 it described how Marat's eyes 'resembled those of the tyger cat', with 'a kind of ferociousness in his looks that corresponded with the savage fierceness of that animal'.[81] In *The Prelude* X. 82, Wordsworth too described Paris at this time as 'a wood where tigers roam'. Blake's characterisations of animals, like Southey's, range from those that are 'sympathetic' to the animal to those that are distinctly 'unsympathetic'; but our modern preoccupation with the question of the author's sympathy can put us at odds with Blake's own symbolic agenda. In this system the larger beasts and birds of prey, such as lions, tigers, wolves, and eagles, generally stand for the natural energies, which may be righteously destructive, with the Tiger symbolising Wrath; and while in the *Songs of Innocence* these animal energies

---

[81] See Martin K. Nurmi, 'Blake's Revisions of "The Tyger",' in Bottrall, pp. 198–217 (p. 200); and Ronald Paulson, 'Blake's Revolutionary Tiger', in *William Blake's Songs of Innocence and of Experience,* ed by Harold Bloom (New York: Chelsea House, 1987) pp. 123–32 (p. 124).

may need to be tamed, in the *Songs of Experience* their destructive instincts often provide a healthy, cleansing force.[82]

It has been suggested that Blake's presentation of 'The Tyger' was conditioned by Edmund Burke's *Reflections on the Revolution in France* (1790), and he may also have used Burke's *Philosophical Inquiry into the Origin of our Ideas of the Sublime and Beautiful* (1757), which categorizes the tiger – along with Leviathan the sea monster or whale, the bull, and the wild ass – as 'sublime' creatures.[83] Burke in the *Inquiry* also demonstrates how a change of context can transform an animal from beautiful or merely innocent, to terrible and sublime:

> Let us look at another strong animal in the two distinct lights in which we may consider him. The horse in the light of an useful beast, fit for the plough, the road, the draft, in every social useful light the horse has nothing of the sublime; but is it thus that we are affected with him, *whose neck is clothed with thunder, the glory of whose nostrils is terrible, who swalloweth the ground with fierceness and rage, neither believeth that it is the sound of the trumpet?* In this description the useful character of the horse entirely disappears, and the terrible and the sublime blaze out together. [84]

This dichotomy between beautiful and innocent animals (which are to be pitied or cared for by human beings) and those that are sublime in their effects upon humanity, wreaking terror and destruction, is particularly pointed and prevalent in the Romantic period – both in its written culture, and in visual art such as the paintings of George Stubbs, Eugène Delacroix and Jean-Louis Géricault.[85] Byron's *Mazeppa,* for example, portrays the horse in strongly-contrasted sublime and beautiful lights: the wild horse to which the young Mazeppa is bound naked is terrifying and destructive, but magnificent in its freedom from humankind, while the old war-horse for which the aged Mazeppa cares tenderly in the framing narrative reflects the horse in the 'useful social light' of Burke's description. In this poem humankind, too, takes on contrasting 'sublime' and 'beautiful' roles, symbolized by Mazeppa's own change from an impetuous youth, full of defiance and sexual energy, to the aged, patient and loyal hetman: 'For danger levels man and brute, / And all are fellows in their need' (51–2).

In Burke, the difference between 'sublime' and 'beautiful' types of animal – like his distinction between the sublime and beautiful in general – is heavily gendered. The beautiful animal may itself have feminine characteristics

---

[82] See F. W. Bateson in his *Selected Poems of William Blake* (London: Heineman, 1957, repr. 1981) p. 117.

[83] See Paulson, p. 125.

[84] Edmund Burke, *A Philosophical Enquiry into the Origin of our Ideas of the Sublime and Beautiful,* ed by J. T. Boulton (London: Routledge and Kegan Paul, 1958) pp. 65–6. Burke is quoting *Job* 29. 19.

[85] Kenneth Clark comments that in the Romantic period 'Man in his relationship with animals began to sympathize with the ferocity, the cruelty even, that he had previously dreaded and opposed. ... [A]t feeding time in the Paris Zoo he [Delacroix] was, he tells us, "pénétré de bonheur".' See *Animals and Men: Their Relationship as Reflected in Western Art from Prehistory to the Present Day* (London: Thames and Hudson, 1977) p. 41.

(particularly a curved form – Burke's example is a dove-like bird – or a gentle, delicate nature) or, as in the examples from Shelley's letters and Byron's *Mazeppa* quoted above, it may draw out feminine characteristics in its human keeper: either erotic ones, or because pets require maternal care and nurturing. Byron points this out in Canto X of *Don Juan*, when he describes Juan with his pets – with a hint that Juan, like his animal companions, is beginning to suffer from being kept in what is in effect captivity, as Catherine the Great's 'man-mistress':

> A bull-dog, and a bull-finch, and an ermine,
>   All private favourites of Don Juan; for
> (Let deeper sages the true cause determine)
>   He had a kind of inclination, or
> Weakness, for what most people deem mere vermin –
>   Live animals: an old maid of threescore
> For cats and birds more penchant ne'er displayed,
> Although he was not old, nor even a maid; –
>
> The animals aforesaid occupied
>   Their station: there were valets, secretaries,
> In other vehicles.
> (X. 393–403)

Byron learnt from the great eighteenth-century French naturalist Buffon, and from Rousseau, the idea that animals under domestication 'degenerated': both from the effects of inbreeding, and because they are, in effect, slaves – like the

>   gazelles, and cats,
> And dwarfs, and blacks, and such like things, that gain
>   Their bread as ministers and favourites – (that's
> To say, by degradation)
> (III. 540–4)

at Juan's and Haidée's sybaritic feast in Canto II of *Don Juan*.

Concern for animals generally was presented by British government ministers as being unmanly and a distraction from the real issues during the war with France; and George Canning's and John Hookham Frere's 'New Morality' (in *The Anti-Jacobin* in 1798) mocked even women who showed excessive sensibility to animals, for their lack of patriotism:

> Taught by nice scale to mete her feelings strong,
> False by degrees, and exquisitely wrong;
> For the crushed beetle first – the widowed dove,
> And all the warbled sorrows of the grove; –
> Next for poor suff'ring guilt; – and last of all,
> For parents, friends, a king and country's fall.

On the other hand, the Newfoundland dog – a very large, water-loving hound with a shaggy black and white coat – was perceived as a notably masculine animal, with sturdy, dependable characteristics. According to Sydenham Edwards, in his *Cynographia Britannica* (1800), 'The size, sagacity, and well known fidelity of these [dogs], deservedly entitle them to the most distinguished rank of all the canine race. ... In swimming and diving, few equal, none excel them, and as their docility is so great, they most readily learn to fetch and carry small burdens in their mouths'.[86] Originating from a part of North America that was still loyal to 'king and country', they could be adopted as particularly British and patriotic. One of the most famous children's books of this period, Thomas Day's *Sandford and Merton* (1783–89) features 'Caesar': 'a large Newfoundland dog, equally famous for his good-nature and his love of water.'[87] Likewise, the behaviour of 'Victor', as reported by Southey in his *Omniana or Horae Otiosores* (1813) was of the kind expected from the reputation of this breed, and was rewarded with a part in a quintessentially British feast:

> There was a Newfoundland dog on board the *Bellona* last war, who kept the deck during the Battle of Copenhagen, running backward and forward with so brave an anger, that he became a greater favourite with the men than ever. When the ship was paid off after the Peace of Amiens, the sailors had a parting dinner on shore. Victor was placed in the chair, and fed with roast beef and plum pudding, and the bill was made out in Victor's name.[88]

Jane Austen in *Northanger Abbey* provides Henry Tilney in his bachelor home at Woodston with 'a large Newfoundland puppy', and its presence is part of a scene which impresses Catherine Morland with Henry's spirited masculinity and suitability as a husband.[89] Byron's fondness for Newfoundlands must also be seen as culturally mediated, and he seems to have been well aware of the potential of his pets' characteristics to express and answer in varying ways his own emotional and representational needs: supplementing aspects of his character or life which he perceived as lacking, or emphasizing traits of which he was proud, such as his swimming abilities.[90]

In 1808, Byron's youthful defiance of the world was expressed – in the form of the 'Inscription' – by his theriophilic preference for dogs over human beings; and in 1824, during his last months in Greece, there was another Newfoundland dog at hand to help represent his middle-aged disillusionment with humanity [see Figure 5]. The artillery officer William Parry describes, in his account of *The Last Days of Lord Byron* (p. 75), how this second Newfoundland, Lyon (a gift from a retired naval officer), provided Byron with a refuge from the barbarous human reality of Missolonghi, and ascribes to

---

[86] Sydenham Edwards, *Cynographia Britannica* (London: printed for the author, 1800) entry for 'Canis Natator' (pages un-numbered after the Introduction).

[87] The edition cited here is Thomas Day, *The History of Sandford and Merton* (London: J. Walker, 1808) p. 228.

[88] See review quoting Southey's item on 'Dogs at Court' in *The Gentleman's Magazine and Historical Chronicle*, LXXXIII. 1, June 1813 (London: Nichols, Son and Bentley) p. 555.

[89] Jane Austen, *Northanger Abbey* ed by Anne Henry Ehrenpreis (Harmondsworth: Penguin, 1972) p. 213.

[90] See Edwards (p. 5): 'The Dog may be considered as, not only the intelligent, courageous, and humble companion of man, he is often a true type of his mind and disposition.'

Byron a tone and a message very little changed from that of sixteen years
before:

> With Lyon Lord Byron was accustomed not only to associate, but to
> commune very much, and very often. His most usual phrase was,
> 'Lyon, you are no rogue, Lyon;' or 'Lyon,' his Lordship would say.
> 'thou art an honest fellow, Lyon.' The dog's eyes sparkled, and his
> tail swept the floor, as he sat with his haunches upon the ground.
> 'Thou art more faithful than men, Lyon; I trust thee more.' Lyon
> sprang up, and barked and bounded round his master, as much as to
> say, 'You may trust me, I will watch you actively on every side.'
> 'Lyon, I love thee, thou art my faithful dog!' and Lyon jumped and
> kissed his master's hand, as an acknowledgement of his homage. In
> this sort of mingled talk and gambol, Lord Byron passed a good deal
> of time, and seemed more contented, more calmly self-satisfied, on
> such occasions, than almost on any other. In conversation and in
> company he was animated and brilliant; but with Lyon and in
> stillness he was pleased and perfectly happy.

It is impossible to tell whether Byron really spoke to the dog in these terms,
for the image of the poet as misanthropic dog-lover was, by the time Parry was
writing, very firmly established, and Parry's image of Byron – like so much
Byronism – cannot now be reliably separated out into those elements which
arose from first-hand observation of the man, and those which could have been
formulated from a reading of his works. An obituarist in *The London
Magazine* (probably Leigh Hunt) drew attention to this continuing element in
Byron's reputation, by recounting an incident which was presented as 'typical'
of the man and his verse:

> One day when Fletcher, his valet, was cheapening some monkeys,
> which he thought exorbitantly dear, and refused to purchase without
> abatement, his master said to him, 'Buy them, buy them, Fletcher; I
> like them better than men – they amuse me and never plague me.' In
> the same spirit is his epitaph on his Newfoundland Dog – a spirit
> partly affected and partly genuine. [91]

Lyon did, however, indeed remain faithful. He accompanied his master's body
by ship to England, arriving in July 1824. But he pined and soon became ill
and was dead himself within a year. The task of burying him fell to Hobhouse,
into whose care he had been placed by Byron's sister Augusta. 'Poor fellow –
he is to be buried under the willow-tree near the water at Whitton,' Hobhouse
recorded. As far as we know, however, no monument or inscription was raised
to mark the grave.[92]

---

[91] *The London Magazine* X (2 October 1824, London: Taylor & Hessey) p. 23.

[92] Hobhouse's diary for 5 June 1825: Broughton Papers, BL Add. MSs. 56549 131r. I am
grateful to Dr Peter Cochran for this and other references to Hobhouse's diaries, and for
allowing me to quote from his transcriptions.

Chapter 2

Children's Animals:
Locke, Rousseau, Coleridge and the
Instruction/Imagination Debate

Jerome McGann's critique of the 'Romantic Ideology' points out that many poetic and critical Romantic self-representations were still dominant, largely unrecognized and unchallenged, in the academy at the end of the twentieth century.[1] A forceful example of McGann's paradigm in action can be found in the way in which literature for children of the Romantic age was predominantly mapped in the modern period as a dichotomy which was specifically based on the views and statements of particular (male) Romantic poets and their circle. When F. J. Harvey Darton and others first began an academic study of Romantic-period children's writing in the 1930s they conceived it in the light of a contrast in value between instruction and imagination, knowledge and romance, fact and fiction: concepts familiar precisely because they were powerfully propounded in some of the poetry of the Romantic period that has continued to be most widely read.[2] Commentary upon education and childhood as a struggle between reason and fantasy, learning and forgetting, is of central importance to poems ranging from Blake's *Songs of Innocence* and *Songs of Experience,* and Wordsworth's 'Intimations of Immortality' Ode, 'Anecdote for Fathers', 'We are Seven' and Book Five of *The Prelude,* to Coleridge's 'Frost at Midnight', 'The Nightingale' and 'Dejection: an Ode'.

It is also from the Coleridge/Wordsworth circle that perhaps the best-known prose exposition of this ideology comes, in the form of Charles Lamb's letter to Coleridge of 23 October 1802, on the subject of buying books for six-year-old Hartley Coleridge. Here the instruction/imagination debate is further polarized into new and old books for children – 'bad' reading and 'good':

> "Goody Two-Shoes" is almost out of print. Mrs Barbauld['s] stuff has banished all the old classics of the nursery; & the Shopman at Newbery's hardly deign'd to reach them off an old exploded corner of a shelf, when Mary ask'd for them. Mrs B's & Mrs Trimmer's nonsense lay in piles about. Knowledge, insignificant & vapid as Mrs B's books convey, it seems, must come to the child in the *shape of knowledge,* & his empty noddle must be turned with conceit of its own powers, when he has learned that a Horse is an Animal, & Billy is better than a Horse, & such like: instead of that beautiful Interest in wild tales, which made the child a man, while all the time he suspected

---

[1]   See McGann, *The Romantic Ideology,* in particular p. 1: 'The ground thesis of this study is that the scholarship and criticism of Romanticism and its works are dominated by a Romantic Ideology, by an uncritical absorption in Romanticism's own self-representations.'

himself to be no bigger than a child. Science has succeeded to Poetry no less in the little walks of Children than with Men. – Is there no possibility of averting this sore evil? Think what you would have been now, if instead of being fed with Tales and old wives fables in childhood, you had been crammed with Geography and Natural History? Damn them. I mean the cursed Barbauld Crew, those Blights and Blasts of all that is Human in man & child.[3]

Adherence to this Romantic dichotomy as a critical position was still being demonstrated as late as 1984 by Geoffrey Summerfield in his *Fantasy and Reason: Children's Literature in the Eighteenth Century*; and although it has been challenged in the last decade by feminist and other studies of writing for children, certain recent re-readings of the views of the Coleridge/ Wordsworth circle are still making use of this paradigm, even if only to argue against its orthodoxy. Alan Richardson, for example, reverses the polarity of Lamb's good and bad readings by finding 'a subtly conservative impetus behind the Romantic defense of fantasy'. He characterizes Wordsworth's criticisms of innovatory educational schemes in book five of *The Prelude* as 'directed against the rational school of educators and writers for children' and so reads Wordsworth's own comments, rather than the work he comments on, as politically reactionary and therefore 'bad'.[4]

This chapter proposes an approach to late eighteenth-century writing for children which does not so obviously privilege the self-representations of the Romantic poets. It explores some of the great wealth of animal material which exists in children's literature in this period as a means of focusing on currents of development which predated and helped to form the imagination/instruction debate, and also ranged beyond it. It too distinguishes a pair of strongly contrasting traditions, illustrating a significant dichotomy in cultural discourse:

---

[2]   F. J. Harvey Darton, *Children's Books in England: Five Centuries of Social Life,* 3rd edn, rev. by Brian Alderson (Cambridge University Press, 1982).

[3]   *The Letters of Charles and Mary Anne Lamb,* ed by Edwin W. Marrs Jr, 3 vols (Ithaca: Cornell University Press, 1976) II. 81–2. Lamb accurately satirizes 'The Art of Distinguishing', an item presented as a discussion between a child and his father in Dr John Aikin's and Anna Laetitia Barbauld's *Evenings at Home; or, The Juvenile Budget Opened,* 3 vols (London: J. Johnson, 1792–96) II. 121. Other Romantic-period criticisms of didactic literature for children can be found, for example, in the work of Coleridge's early mentor Dr Thomas Beddoes, who in *Hygeia* (1802–03) denominated 'the juvenile library' a 'repository of poisons': see Roy Porter, *Doctor of Society: Thomas Beddoes and the Sick Trade in Late-Enlightenment England* (London and New York: Routledge, 1992) p. 68, and William Godwin, who strongly expressed similar views despite contributing widely to such literature himself: see William St Clair, 'William Godwin as Children's Bookseller' in *Children and their Books: a Celebration of the Work of Iona and Peter Opie,* ed by Gillian Avery and Julia Briggs (Oxford: Clarendon Press, 1989) pp. 165–80. More than fifty years later, Charles Dickens was still linking 'old and new' to the difference between imagination and instruction: asking his readers to imagine the contrast between 'the fairy literature of our childhood' and 'a total abstinence edition of Robinson Crusoe, with the rum left out ... [or] a Peace edition, with the gunpowder left out, and the rum left in': see 'Frauds on the Fairies', *Household Words* VIII. 184 (October 1853) pp. 97–100.

[4]   Alan Richardson, 'Wordsworth, Fairy Tales and the Politics of Children's Reading', in *Romanticism and Children's Literature in Nineteenth-Century England,* ed by James Holt McGavran Jr (Athens and London: University of Georgia Press, 1991) pp. 34–53 (p. 43).

but these are traditions which, although they throw light on the familiar poetic distinctions, do not necessarily conform to the parameters set by the Romantic voices which have been dominant for so long.

This reading sees Lamb's opposition of 'science' and 'poetry' in his letter to Coleridge as mirroring the debate which was provoked when, from the early 1760s, the colossal authority of John Locke in educational matters was challenged by the ideas of Jean-Jacques Rousseau. It unpicks Lamb's bundling together of 'Mrs B[arbauld]'s and Mrs Trimmer's nonsense' by demonstrating fundamental differences between these two women authors, in terms of their allegiance to Rousseau and to Locke respectively. It pursues through writing for children the difference between Locke's presentation of the child's mind as a *tabula rasa* or blank page, which must be written on by experience before it can be useful, and Rousseau's notion of the child as in some ways superior to the adult: as one who possesses qualities of innocence and truth-to-nature, which must be lost because of the injurious effects of society as adulthood approaches. 'Everything is good as it comes from the hands of the Author of Nature', Rousseau wrote: 'but everything degenerates in the hands of man' *(Emile,* p. 1).

Using animals

The association of children and animals is often taken for granted, and seems to be of long-standing. Educationalists may claim that children have a natural affinity to animals, or that children tend to identify with the playful, instinctive nature of animals. Anthropologists have ascribed the link to questions of self-definition: pointing out that in every society children have to learn how to distinguish 'Self' from 'Other', including separating kin and friends from strangers and enemies, and 'people' from 'not people' (usually animals). A psychoanalytical approach, such as that of Bruno Bettelheim, suggests that children's belief in animism makes them endow animals with a spirit like their own, and accounts for animal characters in fairy stories in terms of their ability to facilitate, for example, children's coming to terms with their own sexuality.[5] The use of animal material to educate children goes back at least to Aesop in the sixth century BC, and versions of Aesop's fables including those in chapbooks, drawn from the versions of Erasmus (1513), Robert Henryson (1570) and La Fontaine (1651), were frequently used to teach children to read. It was, therefore, not surprising that both Locke and Rousseau should make significant use of animal metaphors to elucidate their ideas, and that the difference between their attitudes should become particularly pointed in the context of material about animals. This is true not only of their own work but also of the educational writing of those who followed them, since animal subjects were at least as prevalent in late eighteenth-century children's literature as they are in modern books, films and other products marketed to children. In Aikin and

---

[5] See Baker, pp. 80–81 and 120–161, citing Richard Tapper, 'Animality, humanity, morality, society', in *What is an Animal?* ed by Tim Ingold, and Bruno Bettelheim, *The Uses of Enchantment: The Meaning and Importance of Fairy Tales* (London: Thames and Hudson, 1976).

Barbauld's highly influential *Evenings at Home,* for instance, of twenty-one items in volume one (1792) ten are about animals; while in volume two (1793) seven out of fifteen items concern them. The work on which Sarah Trimmer founded her right to be regarded as the arbiter and 'Guardian' of children's literature – *Fabulous Histories* (1786: later known as *The History of the Robins)* – is entirely concerned with animal themes.[6] Thomas Day's hugely successful *History of Sandford and Merton* (1783–89) devotes nearly a quarter of its pages to discussing animal topics. The first three chapters of Mary Wollstonecraft's *Original Stories from Real Life* (1788) are dedicated to a discussion of the proper treatment of animals; and *Goody Two-Shoes* (1765) – that text which assumed an almost legendary status in the minds of some of those who read it as children – goes out of its way to show that kindness to animals was one of the main reasons for Margery's success in life.[7]

The animal material in this writing is strongly inter-textual and referential, and many motifs and themes appear time after time. Little Frederick in *The History of the Robins,* for example (p. 78) learns one of Mrs Barbauld's *Hymns in Prose for Children* (1781) by heart; while Mary and Caroline in Wollstonecraft's *Original Stories* read Mrs Trimmer's work with pleasure (p. 11). The rooks which occur briefly in *Goody Two-Shoes* as an early-rising example for gentlemen (p. 24) recur as an item in their own right (still as an illustration of community spirit) in *Evenings at Home* (I. 76). The oft-repeated advice to children to put themselves in an animal's place if they are tempted to torment it, becomes the basis for an entire and very popular genre when Dorothy Kilner's *Life and Perambulation of a Mouse* (1783) ushered in the kind of work where the story is recounted as if by the animal itself.[8]

Locke's linking of animals and children is a prelude to emphasizing the need to impose on children a strict programme of training:

> Why must he at seven, fourteen or twenty years old lose the privilege which the parents' indulgence till then so largely allowed him? try it on a dog or a horse or any other creature, and see whether the ill and resty tricks they have learned when young are easily to be mended when they are knit; and yet none of these creatures are half so wilful and proud, or half so desirous to be masters of themselves and others,

---

[6]   Sarah Trimmer was the editor of *The Guardian of Education* which offered parents guidance on educational matters between 1802 and 1806, 5 vols (London: Hatchard, 1802–06), and of *Fabulous Histories* (1786). The edition cited here is *The History of the Robins, with Twenty-four Illustrations from Drawings by Harrison Weir* (London: Griffith and Farran, n.d.).

[7]   Mary Wollstonecraft, *Original Stories from Real Life 1791,* facs. of 2nd edn (Oxford: Woodstock Books, 1990); Anon., *Goody Two-Shoes: A Facsimile Reproduction of the Edition of 1766,* ed by Charles Welsh (London: Griffith & Farran, 1881).

[8]   Dorothy Kilner, *The Life and Perambulation of a Mouse,* 2 vols (London: John Marshall, 1785). A particular version of the conceit of putting yourself in the place of the animal is provided by stories where a human soul finds itself in an animal's body. An early example is in Catherine Jemmat's *Miscellanies in Prose and Verse* (London, printed for the author, 1766) pp. 133–9, where a country gentleman's son undergoes various indignities and agonies as a result of human cruelty to him in the successive forms of a dog, bullfinch, cockchafer beetle and earthworm. Barbauld's poem 'The Mouse's Petition to Dr Priestley' (1773) and Aikin's and Barbauld's 'The Transmigrations of Indur' in *Evenings at Home* II. 1–34 have the same theme.

as man. ... We are generally wise enough to begin with them when they are very young and discipline betimes those other creatures we would make useful and good for somewhat. They are only our own children that we neglect on this point; and having made them ill children, we foolishly expect they should be good men.[9]

Rousseau also links animals and children, but sees the training of both – and of plants – as a 'mutilating', 'disfiguring' process. Man, he claims (p. 1),

mingles and confounds the climates, the elements, the seasons; he mutilates his dog, his horse, and his slave; he overturns everything, disfigures everything; he loves deformity, monsters; he will have nothing as nature made it, not even man; like a saddle-horse, man must be trained for man's service – he must be made over according to his fancy, like a tree in his garden.

Rousseau's plan for Emile's education consists, therefore, in 'well-regulated liberty'; 'not in gaining time, but in losing it'; 'doing nothing and allowing nothing to be done', so that the pupil is brought 'sound and robust to the age of twelve years without his being able to distinguish his right hand from his left' (p. 59).

The contrasts between Lockean and Rousseauian views became a notable source of tension in education at the end of the century, in the context of new political, religious and revolutionary ideas which broke down consensus and polarized loyalties. The fact that the debate could be seen as an opposition between 'English' and 'French' thinking added to its force. A division arose among educators – concerned as they were with the foundations of society – between those anxious to introduce new ideas of republican citizenship into the education of the next generation, and those battling against what Trimmer saw as 'in respect to *Religion* and *Morals,* a visionary, fallacious and dangerous' system *(Guardian,* I. 184). Those who have attempted to produce a Rousseauian pupil have, she adds *(Guardian,* I. 381), 'been disappointed in their expectations, and have only destroyed what was good, and exhibited to the world their own folly, in adopting a system so contrary to right Reason and the precepts of Christianity.'

This fierce debate was conducted directly, through authors' prefaces and correspondence, as an issue affecting children and their reading, but also more powerfully by indirect means, through writers' choice of subject-matter and its treatment.[10] Animal material in particular was in the forefront of the debate, since the ways in which children were defined in an adult world were often discussed in these texts through animal paradigms, while relationships with pets and other animals provided a way for children to explore their connection

---

[9] John Locke, *Some Thoughts Concerning Education,* ed by F. W. Garforth (London: Heinemann, 1964) pp. 42–5.

[10] See, for example, Trimmer in *Guardian* II. 407: 'The utmost circumspection is therefore requisite in making a proper selection; and children should not be permitted to make their own choice, or to read any books that may accidentally be thrown in their way, or offered to their perusal'. See also Wollstonecraft, in the Preface and Introduction to *Original Stories,* and Day in the Author's Preface to *Sandford and Merton.*

with the rest of society.[11] This becomes particularly explicit in a book such as Trimmer's *History of the Robins,* which alternates between twin animal themes: one, concerned with how children should be brought up to treat animals – 'neither spoil[ing] them by indulgence nor injur[ing] them by tyranny' (p. 140) – and the other using a story of animal (or rather, bird) life as a fable to teach children how to behave correctly in life in society at large. Along with Kilner, Trimmer was, indeed, the first to use this winning formula for children's fiction, which was still being successfully deployed a hundred years later in works such as Anna Sewell's *Black Beauty* (1878).[12] The names of Trimmer's young robins (Dicksy, Pecksy and Flapsy) also suggest that they are the ancestors of another famous set of children's animals: namely Flopsy, Mopsy and Cottontail in Beatrix Potter's *Peter Rabbit.*

Locke and his legacy

Teaching children to be kind to animals was an explicit and substantial part of Locke's method in *Some Thoughts Concerning Education* (1693). He was plainly distressed by children's tendency to '*torment* and treat very roughly young Birds, Butterflies and such other poor Animals which fall into their Hands, and that with a seeming kind of Pleasure' (p. 153). He saw this tendency as 'a foreign and introduced Disposition, an Habit borrowed from Custom and Conversation', and advocated that 'it should be watched in them, and if they incline to any such cruelty they should be taught the contrary usage.'   The training of children to be kind to animals was not, however, an end in itself, for

> they who delight in the Suffering and Destruction of inferior Creatures will not be very apt to be very compassionate or benign to those of their own kind. Our practice takes Notice of this in the Exclusion of *Butchers* from Juries of Life and Death. (p. 153)[13]

These sentiments, drawn from Pythagoras in the classical period and Christian thinking such as that of Thomas Aquinas, found expression in many media in the eighteenth century, including William Hogarth's widely-

---

[11] Locke encouraged the keeping of pet animals, as a way of teaching children (especially girls) the care of and responsibility for dependants: 'a Mother I knew, was wont always to indulge her Daughters when any of them desired Dogs, Squirrels, Birds or any other such things as young Girls use to be delighted with; but then, when they had them, they must be sure to keep them well and look diligently after them, that they wanted for nothing and were not ill-used. ... And indeed, I think People should be accustomed, from their Cradles, to be tender to all sensible Creatures, and to spoil or *waste* nothing at all' (p. 154). See Figure 6.

[12] See Margaret Blount, *Animal Land, The Creatures of Children's Fiction* (London: Hutchinson, 1974) p. 44.

[13] Butchers were apparently still excluded from murder trial juries in the Romantic period: Shelley wrote in 'On the Game Laws': 'the authors of our common law forbid butchers to decide as jurymen on the life of a man because they are familiar, however, innocently, with the death of beasts.' *The Prose Works of Percy Bysshe Shelley,* 2 vols, ed by E. B. Murray (Oxford: Clarendon Press, 1993) I. 281.

disseminated engravings of 1750–1751 on *The Four Stages of Cruelty*.[14] These show a charity boy, Tom Nero, first as the tormentor of animals; next as a coachman beating his horse; then as the murderer of his pregnant mistress, and finally as a hanged corpse being dissected by surgeons, with a dog taking its revenge by devouring one of Tom's discarded organs on the floor (see Figures 7 and 8). In Trimmer's *History of the Robins*, the cruel boy Edward Jenkins, who starts by stealing birds' nests, goes on to 'gratify his malignant disposition on his schoolfellows,' and ends by being killed by a horse that he had been inhumanely beating (p. 139). In Wollstonecraft's *Original Stories*, Mrs Mason recounts (p. 18) how a man who used to let guinea-pigs roll off the roof to see if the fall would kill them (it did), also neglected to educate his children and 'taught them to be cruel while he tormented them', with the consequence 'that they neglected him when he was old and feeble; and he died in a ditch.'

This direct social application of kindness is reinforced by Locke, moreover (p. 156), when he immediately follows his advice about teaching children to be kind to animals with instructions about how tutors should 'accustom [children] to civility in their language and deportment towards their inferiors and the meaner sort of people, particularly servants'. The latter, Locke says, are often treated by gentlemen's children 'as if they were of another race and species beneath them'. There is also an additional benefit for these children in behaving with a 'courteous, affable carriage towards the lower ranks of men,' for 'no part of their superiority will be thereby lost, but the distinction increased and their authority strengthened.'

In other words, the injunction to be kind to animals, like that of being kind to servants and other inferiors, is a part of teaching the (essentially upper-class, male) child its place in the social hierarchy, where beings of all kinds – from those of other species and races, to those of other classes and sexes – are ranged according to the will of God.[15] This aspect of the 'kindness to animals' theme was particularly emphasized by conservative educators such as Trimmer and Hannah More at the end of the eighteenth century. Writing, not as Locke was, for 'gentlemen's sons', but for children of both sexes of the 'middling sort' and the 'lower orders', they used the difference between humans and animals as a way of teaching children the difference between social classes, and as a source of injunctions to those in all ranks of society to be satisfied with their lot.[16] In this way they reflected the fact that literature for children –

---

[14] Hogarth wrote in his *Autobiographical Notes:* 'The four stages of cruelty were done in hopes of preventing in some degree that cruel treatment of poor Animals which makes the streets of London more disagreeable to the human mind, than any thing what ever, the very describing of which gives pain' (quoted in *Hogarth's Graphic Works: First Complete Edition*, ed by Ronald Paulson, 2 vols (New Haven: Yale University Press, 1965), note to plate 201. Leigh Hunt castigated Hogarth's prints because they 'identified virtue with prosperity': see *The Autobiography of Leigh Hunt*, ed by J. E. Morpurgo (London, Cresset Press, 1949) p. 50.

[15] Locke's *Thoughts* were written originally in the form of letters to his friend Edward Clarke, who asked for advice on the upbringing of his eight-year-old son, and were based on Locke's experience as tutor to the family of the Earl of Shaftesbury.

[16] Hannah More's *Cheap Repository Tracts* were issued from 1792 to 1795 at a rate of three a month and were eventually sold in millions; see Cornelia Meigs and others, *A Critical History of Children's Literature* (New York: Macmillan, 1953) p. 80.

particularly in this period – was preoccupied with class. As Robert Leeson comments:

> through their books [children] shared with their parents the hope of rising and the fear of falling. This was the bourgeois age. Trade and commerce opened the way up. But only a few could reach the pinnacle and live securely on their stock as the old style aristocrat lived on his broad acres. The middle class were conscious always of being 'the middling sort'. The master craftsman stood midway between the merchant and banker and the journeyman.[17]

Trimmer's and More's need to counteract the perceived threat of Jacobinism and deist or atheist propaganda led them to stress the disposition of the current social order, as well as the place of animals in it, as the creation and resolution of a beneficent Providence. Thus Trimmer, in the *History of the Robins*, uses the 'animal' side of her story to reflect a hierarchy of *human* social classes – the young robins respect their eldest brother's primogeniture and are cautioned 'not to make acquaintance with sparrows' (p. 89) – while the 'human' side of the tale emphasizes the dominion of humans over animals and the 'ranks' of different species of animal. Mrs Benson, Trimmer's model mother, teaches her children (p. 137) that 'The world we live in seems to have been principally designed for the use and comfort of mankind, who, by Divine appointment, have dominion over the inferior creatures,' and that God 'made all other living creatures likewise; and appointed them in their different ranks in the creation, that they might form together a community, receiving and conferring reciprocal benefits' (p. 136).

The human class system is further used to illustrate a model relationship between humans and (domestic) animals, through Trimmer's good Farmer Wilson's description of the way he treats his animals:

> I always consider every beast that works for me as my servant, and entitled to wages; but as beasts cannot use money, I pay them in things of more value to them ... Besides giving them what I call their daily wages, I indulge them with all the comforts I can afford them (p. 116).

Like good servants, the animals reflect the aspirations of their human masters so that, on the Wilsons' farm,

> The pigs themselves had an appearance of neatness, which no one could have expected in such a kind of animals; and though they had not the ingenuity of the Learned Pig, there was really something intelligent in their gruntlings, and a very droll expression in the eyes of some of them. They knew their benefactors, and found means of testifying their joy at seeing them (p. 99).

---

[17] Robert Leeson, *Children's Books and Class Society: Past and Present* (London: Children's Rights Workshop, 1977) p. 23.

Rousseauian relationships

The application of this human or humane system to animals, as promulgated by Locke and his followers, is not found in Rousseau. True, his ideal pupil Emile 'loves neither disturbance nor quarrels, neither among men, nor even among animals. He will never incite two dogs to fight, and will never cause a cat to be pursued by a dog' (p. 226); but that is as far as his advice goes. Indeed, in *A Discourse on the Origin of Inequality,* Rousseau borrows Descartes' famous description in declaring: 'I see nothing in any animal but an ingenious machine, to which nature hath given senses to wind itself up, and to guard itself, to a certain degree, against anything that might tend to disorder or destroy it'.[18] In *Emile,* moreover, when thehee brother – spite of the fool'is pupil in puberty, and when 'what is needed is a new occupation which keeps him in good humour, gives him pleasure, occupies his attention, and keeps him in training – an occupation of which he is passionately fond and in which he is wholly absorbed', the only one which seems to Rousseau to fulfil all these conditions is hunting, which 'toughens the heart as well as the blood; it accustoms us to blood and to cruelty' (p. 243). The almost universal condemnation of cruelty to animals in late eighteenth-century English children's books cannot be said, then, to owe much to Rousseau. Nevertheless, Rousseau's influence made a significant contribution to the way animals were thought about at this time, in terms of the growth in sensibility towards animals' needs and feelings, and to the perception of them as equal – or in some ways even morally superior – to human beings, which found its expression very clearly in some of the writing for children in this period.

The Cartesian idea of the animal as a mere machine had been developed in works such as Julien-Offray de la Mettrie's *l'Homme-machine* (1747–48) to suggest that human beings as well as animals were machine-like, since the soul could be seen as a function of the body in humans no less than in animals. Rousseau likewise perceives humans as well as animals as machines, but with the distinction that Man is to some extent a free agent and is capable of self-improvement. In *A Discourse on the Origin of Inequality*, however, this distinction is presented as being the reverse of beneficial to humans:

> It would be melancholy, were we forced to admit that this most distinctive and almost unlimited faculty is the source of all human misfortunes; that it is this which, in time, draws man out of his original state, in which he would have spent all his days insensibly in peace and innocence; that it is this faculty which, successively producing in different ages his discoveries and his errors, his vices and his virtues, makes him at length a tyrant over both himself and nature. *(Social Contract,* p. 54)

Moreover, qualities such as compassion and pity, which are absolutely necessary for 'creatures so weak and subject to so many evils as we certainly are', come naturally, 'before reflection', and are therefore more often found in

---

[18] Jean-Jacques Rousseau, *The Social Contract and Discourses,* trans. by G. D. H. Cole (London: Dent, 1973) p. 53.

animals (such as horses, which show a reluctance to trample on living bodies) than they are in humans:

> Compassion must, in fact, be the stronger, the more the animal beholding any kind of distress identifies himself with the animal that suffers. Now, it is plain that such an identification must have been more perfect in a state of nature than it is in a state of reason. ... It is philosophy that isolates him [Man] and bids him say, at the sight of the misfortunes of others: 'Perish if you will, I am secure' (p. 68).

In all, Rousseau concludes in this *Discourse*, 'If she [nature] destined man to be healthy, I venture to say that a state of reflection is one contrary to nature and that the man who meditates is a depraved animal' (p. 51).

Such ideas are found in many forms in late-eighteenth century English literature for children. In volume one of *Evenings at Home* (pp. 81–2), for example, Aikin and Barbauld compare the society of rooks to 'those of men in a savage state, such as the communities of the North American Indians. It is a sort of league for mutual aid and defence, but in which every one is left to do as he pleases'; while a dispute between 'Nature' and 'Education' in volume three (pp. 126–8) concurs with Rousseau's strictures on the pruning of trees:

> While *Nature* was feeding her pine with plenty of wholesome juices, *Education* passed a strong rope round its top, and pulling it downwards with all her force, fastened it to the trunk of a neighbouring oak. The pine laboured to ascend, but not being able to surmount the obstacle, it pushed out to one side, and presently became bent like a bow. Still, such was its vigour, that its top, after descending as low as its branches, made a new shoot upward; but its beauty and usefulness were quite destroyed.

Rousseauian ideas find their most direct form in the writing of Thomas Day: a dedicated Rousseauist and member of the Edgeworth circle, who declared to Richard Edgeworth:

> Were all the books in the world to be destroyed, except scientific books ... the second book I should wish to save, after the Bible, would be Rousseau's *Emilius*. It is indeed an extraordinary work – the more I read the more I admire. ... Excellent Rousseau! First of human kind![19]

Day tried in fact to be a practical as well as a theoretical Rousseauian educator. He attempted (and failed) to raise two orphan girls by Rousseauian methods as potential wives for himself, and he died in his early forties after being thrown and kicked by a colt which he was training according to the principle that animals were well-disposed to humans if kindly treated. His adoption of Rousseauian ideas on animals was part of his generally radical political stance, which also found expression in his hostility to the British position in the American conflict, his campaigning for annual parliaments, and in his

---

[19] Quoted by Sir Michael Sadler, *Thomas Day: An English Disciple of Rousseau*, the Rede Lecture, 1928 (Cambridge: Cambridge University Press, 1928) p. 8.

opposition to slavery. He originally made his name in literature with a well-known poem, 'The Dying Negro' which, as an early biographer commented, 'contributed its share in awakening the feelings of the publick to the sufferings of the Negroes in our islands'.[20]

In Day's most famous children's book *Sandford and Merton* his model child, little Harry the farmer's son, is by page seven already giving evidence of his grasp of the benefits of a natural upbringing:

> it is not fit to mind what we live upon, but we should take what we
> can get, and be contented: just as the birds and beasts do, who lodge
> in the open air, and live upon herbs, and drink nothing but water, and
> yet they are strong, and active and healthy.

A little later we meet Keeper and Jowler, two brother dogs, of whom one is brought up in rough hardihood in the country while the other becomes 'sleek and comely' in the city. But, as would be expected according to Rousseauian principles, it is the unrefined country dog that bravely rescues a man from a wolf, while his city-bred sibling skulks in fear. Later, moreover, when the dogs exchange homes their dispositions are soon reversed also. Mr Barlow, Day's model clergyman and tutor, explains that lions and tigers are only fierce when they are hungry, since they have to kill to live:

> When they are not hungry, they seldom meddle with anything, or do
> unnecessary mischief; therefore they are much less cruel than many
> persons I have seen, and even many children, who plague and torment
> animals, without any reasons whatsoever (p. 43).

Day is much preoccupied with the question of the innocence or guilt of ferocious animals: a concern which even rattled the complacency of divines such as Archdeacon William Paley, the author of *Natural Theology* (1802), who believed that all natural phenomena could be explained as evidences of divine beneficence:

> We have dwelt the longer on these considerations, because the subject
> to which they apply, namely, that of animals *devouring* one another,
> forms the chief, if not the only, instance, in the works of the Deity, of
> an economy, stamped by marks of design, in which the character of
> utility can be called in question.[21]

Mr Barlow believes that 'there is no animal that may not be rendered mild and inoffensive by good usage' – even the crocodile – and when Tommy's tame robin is killed by the cat, Mr Barlow 'consoles' him by showing how robins in their turn kill and eat worms (p. 132). Likewise, when Harry expresses pity for

---

[20] Preface to the 1808 edition of *Sandford and Merton*, p. vi.

[21] William Paley, *Natural Theology; or, Evidences of the Existence and Attributes of the Deity. Collected from the Appearances of Nature* (London: Faulder, 1803) p. 516. Donald Worster, in *Nature's Economy: A History of Ecological Ideas*, 2nd edn (Cambridge: Cambridge University Press, 1994) p. 47, considers that 'the most serious challenge' to the naturalists' 'assertion of cosmic benevolence' was the need 'to explain the bloodshed and suffering in the world.'

a whale which is killed by whalers and 'persecuted for the sake of his spoils', Mr Barlow reminds him that 'the whale himself is continually supported by murdering thousands of herrings and other small fish' (p. 215).

One of the liveliest narratives within this picaresque collection is that of the Highlander who has lived among the American Indians, who are 'bred up from their infancy to a life of equal hardiness with the wild animals,' and who are evidently Day's portrayal of Rousseau's 'noble savages':

> I have seen the greatest and most powerful men of my own country; I have seen them adorned with every external circumstance of dress, and pomp, and equipage, to inspire respect; but I never did see any thing which so completely awed the soul, as the angry scowl and fiery glance of a savage American (p. 333).

Perhaps Day's most daring expression of Rousseauian principles in writing for children comes, however, not in *Sandford and Merton* but in *The History of Little Jack* (1788).[22] In this short story a foundling baby is adopted not only by an old soldier, but also in quite a literal way by the old man's nanny-goat, who is as much the child's 'mother' as the soldier is his 'father' (see Figure 9). When Jack first learns to speak, he begins 'to imitate the sounds of his papa the man and his mamma the goat', and he 'grows up with a combination of hardiness, courage and natural abilities imbibed (literally) from the animal world as well as the human one.[23]

The influence of Rousseau on Wollstonecraft's writing for children is complicated by gender issues. While Wollstonecraft, like Day, broadly approved of Rousseau's ideas on individual choice, on the development of both reason and the emotions and on the concept of the social contract, her own ideas on some aspects of both education and political change differ markedly from Rousseau's. She could not concur with his complete differentiation between the sexes in educational matters, and his limitation of the education of Sophy – Emile's female counterpart – to teaching her how to bear with the injustices inflicted on her by her husband. 'The first and most important quality of a woman is gentleness,' Rousseau maintained *(Emile,* p. 270): 'Made to obey a being as imperfect as man, often so full of vices, and always so full of faults, she ought early to learn to suffer even injustice, and to endure the wrongs of a husband without complaint.' *A Vindication of the Rights of Woman* (1792) commits a complete section (twelve pages in the *Collected Works)* specifically to a critique of this aspect of Rousseau's teaching, as part of Wollstonecraft's 'Animadversions on some of the writers who have rendered women objects of pity, bordering on contempt.'[24] 'Where

---

[22] Thomas Day, 'The History of Little Jack' in *The Children's Miscellany* (London: Stockdale, 1804) pp. 1–57.

[23] Day's conception of his tale may owe something to the story of the baby Jupiter being fed on goat's milk while being brought up by shepherds on Crete, after being smuggled away to safety from the threats of his father Saturn. M. Brown's illustration for the story likewise has similarities to Nicolas Poussin's painting *The Nurture of Jupiter* (c. 1635, Dulwich Picture Gallery), which could be seen in a British collection in the eighteenth century.

[24] *The Works of Mary Wollstonecraft,* ed by Janet Todd and Marilyn Butler, 7 vols (London: Edward Pickering, 1989) V. 147.

he should have reasoned he became impassioned,' she says of Rousseau, 'and reflection inflamed his imagination instead of enlightening his understanding ... I war not with his ashes, but with his opinions. I war only with the sensibility that led him to degrade women by making him the slave of love' (*Works,* V.160–1). In terms of political change, too, Wollstonecraft's programme was very different from Rousseau's. Eschewing his radical or revolutionary rethinking of the social order in favour of a more organic development, she offered, in Anne K. Mellor's words (p. 65): 'an alternative program grounded on the trope of the family-politic, on the idea of the nation-state that evolves gradually and rationally under the mutual care and guidance of both father and mother.'

This complex relationship with Rousseau is demonstrated in Wollstonecraft's references to animals, which use Rousseauian tropes and subject-matter to pursue arguments quite different from his. Thus when Wollstonecraft's model teacher of *Original Stories* – the strangely melancholy, cold and forbidding Mrs Mason – is displeased with her charges she comments: 'You are now inferior to the animals that graze on the common; reason only serves to render your folly more conspicuous and inexcusable' (p. 30). In *A Vindication of the Rights of Woman* Wollstonecraft uses the 'noble savage' metaphor specifically to explain why she believes kindness to animals is *not* one of the British 'national virtues':

> Tenderness for their humble dumb domestics, amongst the lower class, is oftener found in a savage than a civilised state. For civilisation prevents that intercourse which creates affection in a rude hut, or a mud hovel, and leads uncultivated minds who are only depraved by the refinements that prevail in society, where they are trodden under foot by the rich, to domineer over them to revenge the insults that they are obliged to bear from their superiors. *(Works,* V. 243)

And although here and in *A Vindication of the Rights of Men* (1790), Wollstonecraft is deeply critical of the class system, it is a conservative preoccupation with the significance and the maintenance of hierarchies, distinctions and ranks which is one of the most characteristic traits of her writing for children.[25] In *Rights of Men (Works,* V. 46) she had maintained that 'Inequality of rank must ever impede the growth of virtue, by vitiating the mind that submits or domineers.' In her work for children, however, as in Trimmer's, human beings' place in the chain of being is stressed time and time again: 'Mary interrupted her to ask, if insects and animals were not inferior to men; Certainly, answered Mrs Mason; and men are inferior to angels' *(Original Stories,* p. 13); and:

> The children eagerly enquired in what manner they were to behave, to prove that they were superior to animals? The answer was short, – be

---

[25] Hannah More is said to have commented when Wollstonecraft's *Rights of Women* appeared: 'Rights of Women! We'll be hearing of the Rights of Children next!' Quoted in Robert Woof, Stephen Hebron and Claire Tomalin, *Hyenas in Petticoats, Mary Wollstonecraft and Mary Shelley* (Kendal: The Wordsworth Trust, 1997) p. 69.

tender-hearted; and let your superior endowments ward off the evils
which they cannot foresee. It is only to animals that children can do
good, men are their superiors (p. 16).

On the other hand, Mrs Mason tells the girls that 'Children are inferior to
servants – who act from the dictates of reason, and whose understandings are
arrived at some degree of maturity' (p. 103); and when Mary protests that she
does not need to be kind to worms because they are 'of little consequence in
the world', her teacher harshly responds: 'You are often troublesome – I am
stronger than you – yet I do not kill you' (p. 4).

Wollstonecraft's ideas about animals and their relationship with humans in
her writing for children are on the whole, therefore, closer to those of Trimmer
and other political conservatives than those of Rousseau, Day or Richard
Edgeworth. In terms of the interchange of feeling between humans and
animals, however, there are significant differences between Trimmer and
Wollstonecraft. Trimmer constantly warns her readers not to become too
attached to animals:

> It is wrong to grieve for the death of animals as we do for the loss of
> our friends, because they certainly are not of as much consequence to
> our happiness; and we are taught to think their sufferings end with
> their lives, as they are not religious beings; and therefore the killing
> them, even in the most barbarous manner, is not like murdering a
> human creature, who is, perhaps, unprepared to give an account of
> himself at the tribunal of Heaven (p. 47).

Wollstonecraft, however, frequently creates situations where an animal
becomes the focus of great sentimentality and pathos. Again, in view of her
resistance to 'sensibility' in both the *Vindications,* it is surprising to find her in
Original Stories invoking her readers' sentimental feelings in sometimes quite
relentless terms. Thus the story of Poor Mad Robin, who begins as a
respectable workman and loses first his living, and next his wife and then his
children, reaches its crescendo only when his faithful dog expires also, having
amply demonstrated the Rousseauian principle that animals are more capable
of compassion than humans:

> Will any one be kind to me! [Poor Robin asks] – you will kill me! – I
> saw not my wife die – No! – they dragged me from her – but I saw
> Jacky and Nancy die – and who pitied me? – but my dog! He turned
> his eyes to the body – I wept with him (p. 26).

The sentiment of even this story can be capped, however, as it is immediately
in the book when one of the children asks, 'Did you hear of anything so cruel?
Yes, answered Mrs Mason, and as we walk home I will relate an instance of
still greater barbarity;' and she proceeds to recount the story of a Bastille
prisoner who had befriended a spider, which is deliberately crushed by a cruel
guard (p. 27).

Such writing about animals reached its apogee after Wollstonecraft's death,
in works such as Augustus Kendall's *Keeper's Travels in Search of his Master*

(1798), and William Hayley's *Ballads* (1805) with its story of Lucy's Fido (see Figure 10).[26] This dog saves Lucy's beloved Edward from death in a crocodile-infested river in India, but in the process is drowned himself. He is commemorated in a statue which (once they are married) is placed in Lucy's and Edward's bedroom:

> The marble Fido in their sight,
>   Enhanc'd their nuptial bliss;
> And Lucy every morn, and night,
>   Gave him a grateful kiss.
> *(Ballads,* p. 12)

Despite his condemnation of Hayley's style as 'forever feeble and forever tame' *(English Bards,* 314). it was only three years after this that Byron was commemorating his own dog Boatswain, also in marble and in equally fulsome terms, in his monumental 'Inscription'. Trimmer, however, would have none of such sentiment, holding that 'to talk of the virtues of the lower animals' was not consistent 'with the order of the creation'. She was still further opposed to the idea that animals might have rights: a notion put forward by Kendall in the Dedication to *Keeper's Travels,* which provoked a devastating review by her in *The Guardian of Education:* 'the levelling system, which includes the RIGHTS OF ANIMALS, is here carried to the most ridiculous extreme.'[27] Happily, however, she concluded,

> the cattle of the field, and the fowls of the air, and all the animal tribes are secured, by their limited faculties, from the influence of their Advocate's sophistical arguments; and kept in their proper stations by the over-ruling power of the Almighty! (I. 400)

Trimmer here appears genuinely concerned about the possibility that arguments for animal rights might give ideas above their station to some potentially upstart beasts. But the essential battle was over the minds of the children: Trimmer called the choice of reading-matter for them a 'momentous concern' *(Guardian,* II. 407), and author after author testified to the profound effect that their childhood reading had had on their view of life. Leigh Hunt, for example *(Autobiography,* p. 51) described how Day's *Sandford and Merton* had

> assisted the cheerfulness I inherited from my father; showed me that circumstances were not to crush a healthy gaiety, or the most masculine self-respect; and helped to supply me with the resolution of standing by a principle, not merely as a point of view of lowly or lofty

---

[26] Augustus Kendall, *Keeper's Travels in Search of his Master* (London: E. Newbery, 1798); William Hayley, *Ballads ... Founded on Anecdotes Relating to Animals; with prints designed and engraved by W. Blake* (Chichester: Richard Phillips, 1805).

[27] 'I cannot help anticipating the time,' Kendall wrote (p.v), 'when men shall acknowledge the RIGHTS; instead of bestowing their COMPASSION upon the creatures whom, with themselves, GOD made, and made to be happy! – If any part of their condition is to be compassionated, – it is that they are liable to the tyranny of man.'

sacrifice, but as a matter of common sense and duty, and a simple co-operation with the elements of natural warfare.

Coleridge and children's reading

Because it forms such a large portion of children's reading, material about animals ought, according to contemporary belief, to show this formative effect particularly strongly. The second part of this chapter therefore addresses the ways in which prevailing attitudes towards animals might be engaged with in the work of a writer who had been exposed to some of this material. Coleridge offers a particularly interesting subject of study in this context for several reasons. First, his recollection of his own childhood reading was preternaturally sharpened by and combined with the acquisition of a close knowledge of contemporary educational theories when the upbringing of his son Hartley became the subject of wide-ranging debate in the Coleridge/Wordsworth circle.[28] Second, Coleridge's poetry and prose writing includes several intense imaginative engagements with animals and some of his most interesting metaphorical self-identifications are with non-human species, especially birds. And last, he articulates his thinking about what it means to be human through brilliant and extensive ruminations in prose about the status of animals vis à vis that of humankind.

Coleridge's early childhood in the 1770s was characterized by voracious reading of all sorts, and he left vivid accounts of consuming (or being consumed by) the *Arabian Nights* and 'all the gilt-cover little books that could be had at that time' from his aunt's shop in Crediton.[29] These gilt covers were the hall mark of the collections of the children's publisher John Newbery, and among the other Newbery publications which Coleridge may have read are *Tommy Trips's History of Birds and Beasts* (1760); *The Newtonian System of Philosophy* (1761); the 'Gift' series, such as *The Valentine's Gift* (1765) and *The Twelfth-Day Gift* (1767) and, almost certainly, in view of his later familiarity with it, Newbery's most famous production *The History of Little Goody Two-Shoes, Otherwise Called, Mrs Margery Two-Shoes* (1765).[30]

Newbery was an energetic, entrepreneurial and very successful publisher, nicknamed 'Jack Whirler' by Samuel Johnson. His books were in many ways apologias for middle-class commercial activity and his motto was 'Trade and Plumb-cake for ever!'. He was also a thorough-going Lockean, and his publications for children combined the promotion of kindness to animals with

---

[28] Coleridge wrote to his wife on 18 September 1798, for example: 'read Edgeworth's Essay on Education – read it heart & soul – & if you approve of the mode, teach Hartley his Letters. I am very desirous, that you should begin to teach him to read – & they point out some easy modes. – J. Wedgewood informed me that the Edgeworths were most miserable when Children, & yet the Father, in his book, is ever vapouring upon their Happiness! – However, there are very good things in the work — and some nonsense!' *Collected Letters*, I. 418.

[29] Letter to Thomas Poole, 9 October 1797, *Collected Letters*, I. 347.

[30] Meigs and others (p. 64) identify 'gilt and embossed paper covers' as 'the mark of Newbery's taste'.

the inculcation of bourgeois values.[31] Lucy Newlyn suggests that the relationship between Coleridge and Wordsworth involved a downplaying of the aspects of their past and experience that they did not share, with a privileging of the Wordsworthian childhood as an 'ideal', and that Hartley's birth provided Coleridge with the opportunity to vicariously create for himself a version of the rural childhood he had himself missed through being sent to school in London.[32] In this climate, elements of Coleridge's childhood reading which fitted in with the Wordsworthian emphasis on 'wildness' (such as the 'wild tales' of the *Arabian Nights* and the chapbook stories) would be stressed, while those that did not conform to the Wordsworthian model (such as the other bourgeois, urban-oriented Newbery books) would be suppressed or ignored.[33] Thus Coleridge claims that it is the 'Faery Tales ... Romances, & Relations of Giants and Magicians, & Genii' that had 'habituated [his mind] *'to the Vast'* and given him 'a love of "'the Great,'' & "'the Whole,'''' but does not mention the Newbery books, such as *Goody Two-Shoes,* in this context, although there is no doubt that he knew them well *(Collected Letters,* I. 354). Following Coleridge's and Wordsworth's own paradigm, the influence on Coleridge's creative work of the *Arabian Nights* and the Faery Tales has been widely acknowledged and explored by the critics, while the influence of the Lockean instructional stories of his own childhood has been largely ignored. The following part of this chapter seeks to show how the books from both traditions combined to form a rich source of material with which Coleridge avidly engaged in verse and in prose.[34]

The association of kindness to animals, entrepreneurism and 'getting on in life' is certainly characteristic of *Goody Two-Shoes,* where Margery in her village school is very specific in teaching her pupils about kindness to animals, uses pet animals as 'monitors' to help children to read, and even risks being branded a witch because her pet crow, dog, pigeon, lark and lamb are taken by the ignorant to be her familiars. Although, when he is writing to Coleridge in 1802, Lamb places *Goody* in firm and positive opposition to the didactic 'nonsense' of Mrs Barbauld and Mrs Trimmer, *Goody's* tone on many moral questions, and certainly on animals, is in fact very similar to the productions of these later writers:

---

[31] Samuel F. Pickering Jr, *John Locke and Children's Books in Eighteenth-Century England* (Knoxville: University of Tennessee Press, 1981) pp. 12–14.

[32] Lucy Newlyn, *Coleridge, Wordsworth and the Language of Allusion* (Oxford: Clarendon Press, 1986) pp. 141–64.

[33] The Coleridge circle's description of the tales as 'wild' echoes Samuel Johnson's in the *Dictionary* (1755), where he defines 'romance' as 'a military fable of the middle ages; a tale of wild adventures in war and love,' and 'romantick' as 'resembling the tales of romances; wild ... improbable, false ... fanciful; full of wild scenery': *A Dictionary of the English Language,* facs. of Knapton edn, 1755 (London: Times, 1983). According to Thomas Percy, Johnson 'attribute[d] to those extravagant fictions that unsettled turn of mind which prevented his ever fixing in any profession,' and Eithne Henson claims he was 'addicted to chivalric romance': see *'The Fictions of Romantick Chivalry': Samuel Johnson and Romance* (Rutherford: Fairleigh Dickinson University Press, 1992) p. 19.

[34] T. S. Eliot's short introduction to Coleridge for a National Portrait Gallery postcard began: 'When five years old had read Arabian Nights'.

> Mrs *Margery,* you must know, was very humane and compassionate;
> and her tenderness extended not only to all Mankind, but even to all
> Animals that are not noxious; as your's ought to do, if you would be
> happy here, and go to Heaven hereafter. These are God Almighty's
> creatures as well as we. He made both them and us; and for wise
> Purposes, best known to Himself, placed them in this world to live
> among us; so that they are our fellow Tenants of the Globe. How then
> can People dare to torture and wantonly destroy God Almighty's
> Creatures? They as well as you are capable of feeling Pain, and of
> receiving Pleasure, and how can you, who want to be made happy
> Yourself, delight in making your fellow Creatures miserable? (p. 68)

Margery acquires her raven from 'wicked boys' who were going 'to throw at
it', and her pigeon from

> naughty Boys, who had taken [it] and tied a string to its Leg, in order
> to let it fly, and draw it back again when they pleased; and in this
> manner they tortured the poor Animal with hopes of Liberty and
> repeated Disappointment (p. 69).

It is this kind of incident that Coleridge drew upon in a satirical political
poem *Parliamentary Oscillators* (1798), which has 'a waggish crew' who
fasten an owl to a duck's back and throw them both into a pond for sport. He
may have been punning on the title of *Goody Two-Shoes* when he commented
that 'moral tales' for children taught not 'goodness, but – if I might venture
such a word – goodyness,' and when he described how his changes to a later
poem about childhood reading had been made because of 'a cowardly fear of
the Goody'.[35] The 1794 Pantisocratic poem 'To a Young Ass, Its Mother
being tethered near it' addresses animal treatment in language relating to the
equality and rights of *human* groups, particularly slaves:

> Poor little Foal of an oppressed Race!
> I love the languid patience of thy face:
> And oft with gentle hand I give thee bread,
> And clap thy ragged Coat and pat thy head. ...
> Do thy prophetic Fears anticipate,
> Meek Child of Misery! thy future fate?
> The starving meal, and all the thousand aches
> 'Which patient Merit of the Unworthy takes'?
> Or is thy sad heart thrilled with filial pain
> To see thy wretched Mother's shortened Chain? ...
> Innocent Foal! thou poor despised Forlorn!
> I hail thee brother – spite of the fool's scorn!
> And fain would take thee with me, in the Dell
> Of Peace and mild Equality to dwell.
> ('Young Ass' 1–4; 9–13; 25–8) [36]

---

[35] *Coleridge's Shakespearean Criticism,* ed by Thomas Middleton Raysor, 2 vols
(London: Constable, 1930) II. 13.

[36] Samuel Taylor Coleridge, *Poetical Works* ed by E. H. Coleridge, 2 vols (Oxford:
Clarendon Press, 1912).

It is difficult to judge the precise tone of this poem, between serious sentimental humanitarianism and burlesque.[37] It shares this uncertainty of approach with Southey's 1790s animal poems, discussed in Chapter One, to which it forms a companion. An early manuscript of the poem (entitled 'Monologue to a Young Jack Ass in Jesus Piece. Its mother near it chained to a log', see *Poetical Works,* ed Coleridge, I. 75n) ends in burlesque, with an introduction to the Pantisocratic Dell

> Where Mirth shall tickle Plenty's ribless side,
> And smiles from Beauty's Lip on sun-beams glide,
> Where Toil shall wed young Health that charming Lass!
> And use his sleek cows for a looking-glass –
> Where Rats shall mess with Terriers hand-in-glove
> And Mice with Pussy's Whiskers sport in Love!

There are references to animals with a similar facetiousness and ambivalence of tone in Coleridge's letter to Francis Wrangham, written the same day:

> If there be any whom I deem worthy of remembrance – I am their Brother. I call even my Cat Sister in the Fraternity of universal Nature. Owls I respect & Jack Asses I love: for Aldermen & Hogs, Bishops and Royston Crows I have not particular partiality – ; they are my Cousins however, at least by Courtesy. But Kings, Wolves, Tygers, Generals, Ministers, and Hyaenas, I renounce them all – or if they *must* be my kinsmen, it shall be in the 50th Remove – May the Almighty Pantisocratizer of souls pantisocratize the Earth, and bless you and   S. T. Coleridge! – *(Collected Letters,* I. 121)

The published version of the poem has an amended, 'serious' ending, but Carl Woodring believes that it preserves the irony and the potentially comic intention of the burlesque lines:

> The poem presents to the reader the irony, comic or not, that the extreme sadness of the ass may come not from its own miserable lot or anticipation of the worsening lot that awaits it, but from pity for its more strictly chained mother, and the related irony that the master of the pair has not learned from his own very similar state of oppression. ... Even when reduced to the inhuman level of a jackass, the man has less pity than the jackass has (p. 71).

Reference to the *Discourse on the Origin of Inequality* demonstrates, however, how the Rousseauian point is being made here that compassion may be *more* natural to animals, and to human beings 'in a state of nature' than it is to urbanized and supposedly civilized humanity. In the following year, Coleridge as political lecturer described how the brutalization of the working class reduced them to something lower than the animals: 'can we wonder that men should want humanity, who want all the circumstances of life that humanize? Can we wonder that with the ignorance of Brutes they should unite

---

[37] Carl R. Woodring, *Politics in the Poetry of Coleridge* (Madison: University of Wisconsin Press, 1961) pp. 60–72.

their ferocity?'[38]   In the poem, therefore, he can be seen pursuing a
Rousseauian, logical and not necessarily ironic or comic argument about the
unwelcome effects of 'civilization' upon his urbanized contemporaries, which
has affinities with Wollstonecraft's presentation of the same theme in *Original
Stories*, and with Coleridge's own intention – current at this time – to retire
from the world into a Pantisocratic retreat.   That Coleridge felt some
commitment to the poem can be demonstrated by the fact that he continued to
publish it in its 'serious' form for many years, despite constant jibes such as
Byron's in *English Bards and Scotch Reviewers* (1809):

> Shall gentle Coleridge pass unnoticed here,
> To turgid ode, and tumid stanza dear? ...
> The bard who soars to elegise an ass
> So well the subject suits his noble mind,
> He brays, the laureat of the long-ear'd kind.
> *(English Bards, 255–64)*

In view of Byron's own somewhat dubious public association with animals,
Coleridge might well have dismissed this as 'the pot calling the kettle black';
but his ambivalence towards his own poem may be indicated by its omission
from *Sibylline Leaves* in 1817.

'The Ancient Mariner'

'To a Young Ass' opens with Lockean attitudes on kindness to animals:

> And oft with gentle hand I give thee bread,
> And clap thy ragged Coat and pat thy head ...
> Poor Ass! thy master should have learnt to show
> Pity ...

It is made complex and more interesting, however, by the overlay of
Rousseauian notions about animals' own ability to feel compassion, and the
political context of animals' and humans' common suffering brought about by
living 'Half famished in a Land of Luxury' (l. 22).   'The Ancient Mariner'
most obviously draws on the other strand of Coleridge's boyhood reading: the
Arabian Nights, and the 'Faery Tales ... Romances, & Relations of Giants and
Magicians & Genii' to which he paid greatest tribute as a source of his
imaginative work. 'The Mariner' was written, however, in 1797–98: just at the
point when Hartley (who was born in September 1796) was old enough for a
debate about his education to begin, and so the animal elements in it reflect
aspects of the children's books of the 1780s and 1790s as well as of
Coleridge's own childhood reading.   The way in which Coleridge himself
connected the animals in contemporary children's books with 'The Mariner' is
illustrated by his remarks from *Table Talk* for 31 March 1832:

---

[38] Coleridge, *Lectures 1795 on Politics and Religion,* ed by Lewis Patton and Peter Mann,
The Collected Works of Samuel Taylor Coleridge, Bollingen Series (London: Routledge &
Kegan Paul, 1971) p. 10.

Mrs Barbauld told me that the only faults she found with the Ancient Mariner were – that it was improbable, and had no moral. As for the probability – to be sure that might admit some question – but I told her that in my judgment the chief fault of the poem was that it had too much moral, and that too openly obtruded on the reader. It ought to have had no more moral than the story of the merchant sitting down to eat dates by the side of a well and throwing the shells aside, and the Genii starting up, and saying he must kill the merchant, because a date shell had put out the eye of the Genii's son.[39]

It seems odd that Barbauld, herself a prolific writer of didactic children's material, should have missed the apparently very obvious moral message, repeated by the poem's narrator almost in the form of a refrain:

> He prayeth well, who loveth well
> Both man and bird and beast.
>
> He prayeth best, who loveth best
> All things both great and small:
> For the dear God, who loveth us,
> He made and loveth all.
> (1798 version, 645–50)

The ballad-like metrical form and simple syntax seem to imitate precisely the tone of didactic literature for children of this period. Barbauld may have felt, however, that the action of the story did not lead inevitably to the conclusion indicated by the moral, and that a lesson out of a contemporary English children's book had been gratuitously tacked on to an *Arabian Nights*-style story. Coleridge, on the other hand, evidently meant by his claim that there was 'too much' moral that he felt the poem suffered from being in this respect too like the sort of children's book he, Wordsworth and Lamb had castigated many years before. The 'Faery Tale' milieu of 'The Mariner' is of exactly the kind driven out of fashion for a while at the turn of the century by books for children like Barbauld's, and around 1803 Barbauld represented a kind of bogey-figure in the letters of the Coleridge circle.[40] Seven years earlier, however, in a *Watchman* review of 1796, Coleridge had acknowledged Barbauld's effectiveness in a role not dissimilar to his own in 'To a Young Ass', when he remarked that 'thanks to Mrs Barbauld ... it has become

---

[39] *Samuel Taylor Coleridge, Table Talk: Recorded by Henry Nelson Coleridge (and John Taylor Coleridge)* vol. I, ed by Carl Woodring, *The Collected Works of Samuel Taylor Coleridge*, Bollingen Series (London: Routledge, 1990) pp. 272–3. In another report, H.N.C. recorded S.T.C. as saying specifically: 'In a work of such pure imagination I ought not to have stopped to give reasons for things, or inculcate humanity to brutes' *(Table Talk*, p. 273, n. 7).

[40] On 8 May 1812 Henry Crabb Robinson recorded: 'Wordsworth is not reconciled to Mrs Barbauld; his chief reproach against her now is her having published pretty editions of Akenside, Collins, etc., with critical prefaces which have the effect of utterly forestalling the natural feeling of young and ingenuous readers': *Henry Crabb Robinson on Books and their Writers*, ed by Edith J. Morley, 3 vols (London: Dent, 1938) I. 74. Barbauld was wrongly supposed to have written an unfavourable review of Lamb's play John Woodvil (see S. T. C.: *Table Talk*, p. 272, n. 6). Coleridge referred to Dr John Aikin, Barbauld's brother, as an 'aching void', and to his son, adopted by Barbauld, as 'a void aching' (Summerfield, p. 264).

universally fashionable to teach lessons of compassion towards animals.'[41]
Moreover, the power of 'The Ancient Mariner' itself should not be
underestimated, at least in the nineteenth century, to preach effectively the
simple moral of kindness to animals. In 1829, for instance, the Cambridge
Union Society debated whether 'Mr Coleridge's Poem of the Ancient Mariner,
or Mr Martin's Act [will] tend most to prevent cruelty to animals?' ('Mr
Coleridge's Poem' just won, by 47 votes to 45).[42] And as late as 1881
Frederick Thrupp, lithograph illustrator of Coleridge's work, was interpreting
the poem in this straightforward spirit in his essay decrying blood-sports, 'The
Antient [sic] Mariner and the Modern Sportsman'.[43]

A satirical and parodic approach to Barbauld and other didactic educators
appears also in a short poem, written at the same time as the first version of
'The Ancient Mariner.' In 'The Raven' (first published in 1798), later sub-
titled 'a Christmas tale, told by a school-boy to his little brothers and sisters',
the child imitates the moralistic tone of such writers while recounting a tale
with a 'moral' which is a deliberate overturning of the sort that their work
might promulgate. It tells its story from the point of view of a Raven, who
plants an acorn and waits for it to grow into an oak-tree. When he and his
mate are nesting in it a Woodman comes and cuts it down, destroying the nest,
mate and nestlings. The tree is made into a ship, and the Raven gets his
revenge when this is wrecked with much loss of human life: 'The Raven was
glad that such fate they did meet, / They had taken his all, and REVENGE IT
WAS SWEET!' (1798 version, 41–2). The Raven and the child raconteur are
united in seeing themselves as the centre of the universe – a trait characteristic
of both real children and real animals – and moreover they 'have a right to
their angry feelings over life's injustices'.[44] Coleridge presents the yearning of
a child, and the animal character which speaks for him, to find fairness and
justice in a real world which is more complex, subtle and amoral than anything
he believed the didactic educators could envisage. Even in this location,
however, Coleridge's ambivalence towards the educators is evident: in the
1817 *Sibylline Leaves* version of the poem two lines were added at the end
which undo the effect of the parody and turn it back into an object lesson for
children: 'We must not think so; but forget and forgive, / And what Heaven
gives life to, we'll still let it live' (45–6). Later still Coleridge added a
manuscript note to the passage claiming that these lines had been:

---

[41] *The Watchman* IX (5 May 1796), in *The Watchman* ed by Lewis Patton, *The Collected
Works of Samuel Taylor Coleridge,* Bollingen Series, (London: Routledge & Kegan Paul,
1970) p. 313. In 1797 Coleridge walked twenty miles to Bristol and back to see Barbauld, and
in March 1798 he described her as 'that great and excellent woman': see *Coleridge, Table
Talk*, p. 272, n. 6.

[42] Cambridge Union Society, *Laws and Transactions of the Union Society,* (Cambridge,
1830) p. 56: topic for debate on 17 February 1829. Richard Martin's Act for the prevention of
cruelty to domestic animals was passed in 1822.

[43] Frederick Thrupp, 'The Antient Mariner and the Modern Sportsman' (London: James
Martin, 1881).

[44] Jeanie Watson, 'Coleridge and the Fairy Tale Controversy,' in *Romanticism and
Children's Literature,* ed by McGavran, pp. 14–33 (p. 29).

Added thro' cowardly fear of the Goody! What a Hollow, where the Heart of Faith ought to be, does it not betray? this alarm concerning Christian morality, that will not permit even a Raven to be a Raven, nor a Fox a Fox, but demands conventicular justice to be inflicted on their unchristian conduct, or at least an antidote to be annexed. *(Poetical Works* ed by E. H. Coleridge, I. 171, n. 40)

As demonstrated by these fluctuations in authorial and editorial intention, and by the existence of both burlesque and serious versions of 'To a Young Ass', Coleridge's attitudes to animal nature and in particular to natural animal bloodthirstiness are complicated by uncertainty, which is sometimes disguised by facetiousness and sometimes allowed to express itself honestly. Reference to the struggles of the anxious educators, attempting to present both animal goodness and animal cruelty within a single coherent moral framework, and to Coleridge's own ambivalence about the nature of animals, throws light on the significance of the non-human creatures of 'The Ancient Mariner'. The albatross and the water-snakes (which may or may not be the same as 'the slimy things' that 'did crawl with legs / Upon the slimy sea') have not only beneficent, but also ambiguous and amoral characteristics, and appear both as examples of the goodness of the natural world and also as the apparent cause of excessive (human) guilt and suffering.[45] But whereas in a didactic genre such as writing for children the need to be morally definite is paramount, Coleridge in this imaginative context is able to use multiple voices and viewpoints and hold these complexities in balance, so that meanings can be fluid and unfixed. Although an emblematic significance or 'moral' seems to be constantly intimated for the albatross and the water-snakes, no single, constant meaning can be reliably allocated to them, and this mystery and unpredictable peculiarity is a major cause of their attractiveness and the power of the poem.

Animals to characterize the self

Also indicative of Coleridge's reaction against the single-minded didacticism of contemporary writing about animals and other natural objects is his statement in an 1802 letter to William Sotheby about the poems of William Bowles:

There reigns throughout ... such a perpetual trick of *moralizing* every thing – which is all very well, occasionally – but never to see or describe any interesting appearance in nature, without connecting it with dim analogies with the moral world, proves Faintness of Impression. Nature has her proper interest; & he will know what it is, who believes & feels, that everything has a life of it's own, & that we are all *one Life*. A Poet's *Heart & Intellect* should be combined, intimately combined & *unified*, with the great appearances in Nature

---

[45] Ian Wylie, in *Young Coleridge and the Philosophers of Nature* (Oxford: Clarendon Press, 1989) pp. 143–62, puts forward a convincing case that the 'slimy things' are the larval stage of the later snakes, and are the supposed result of spontaneous generation awakened by the breeze from the rotting deeps which surround the ship.

& not merely held in solution & loose mixture with them, in the shape
of formal Similes. *(Collected Letters,* II. 459)

Coleridge finds Bowles's way of working not dissimilar to the predominant
methods of children's writing at this time, characterized by Summerfield as the
'jaded traditions ... [of] the fable, which offered examples of virtue and vice in
action; and the emblem, which presented an object for contemplation, the
allegorical or anagogical significance being carefully disentangled and
explained' (p. 233). It is not, however, the principle of observing similarities
between the natural world and the human moral world which Coleridge objects
to, nor the relegation of 'appearances in Nature' to the subordinate role in
comparisons, but the use in verse of superficial likenesses, mechanistically
joined, instead of a recognition of the real unity between the human and the
natural world, which he believes can only be perceived by true poets and
expressed by deep metaphor. Coleridge's own work in both verse and prose
frequently deploys animals in order to make analogies between the natural and
the human world. Thus when his capacity for intense observation is brought to
bear upon the larva of the beetle, which leaves room within its chrysalis for the
development of parts it does not yet have, it is used as a simile first for human
physical development and then for spiritual growth, within Coleridge's
Romantic evolutionary scheme:

> The Larva of the stag-beetle lies in its Chrysalis like an infant in the
> Coffin of an Adult, having left an empty space half the length it
> occupies – and this space is the exact length of the Horn that
> distinguishes the perfect animal, but which, when it constructed it's
> temporary Sarcophagus, was not yet in existence. Do not the Eyes,
> Ears, Lungs of the unborn Babe give notice and furnish proof of a
> transuterine, visible, audible, atmospheric world? ... But likewise –
> alas for the Man, for whom one has not the same evidence of the Fact
> as the other – the Creator has given us supernatural Senses and Sense-
> organs – Ideas, I mean! ... and must not these too infer the existence of
> a world correspondent to them? *(Collected Letters,* VI. 595)[46]

Throughout his writing Coleridge demonstrates a particularly keen interest
in birds, and in a manuscript note in a copy of Gilbert White's works he
describes how he had himself 'made and collected a better table of characters
of Flight and Motion' of birds than White's.[47] His propensity to compare
human and natural objects, moreover, leads him not only to see the natural
world as a source of images which illuminate the human condition, but also to
deploy human artefacts as a means of describing natural phenomena. Thus a
flock of starlings in flight is transformed into a series of abstract objects drawn
from human geometry:

> Starlings in vast Flights, borne along like smoke, mist – like a body
> unindued with voluntary Power – now it shaped itself into a circular

---

[46] *Larva* in Latin means a spectre or ghost.
[47] See George Whalley, *'The Mariner and the Albatross'* in *Coleridge:* The Ancient
Mariner *and Other Poems: A Casebook,* ed by Alun R. Jones and William Tydeman (London:
Macmillan, 6th repr. 1990) p. 181.

area, inclined – now they formed a Square – now a Globe – now from complete Orb into an Ellipse – then oblongated into a Balloon with the Car suspended, now a concave Semicircle; still expanding, or contracting, thinning or condensing, now glimmering and shivering, now thickening, deepening, blackening![48]

Some of the most powerful of Coleridge's references to animals are those in which he uses birds as symbols for aspects of himself, and George Whalley (p. 172) considered the albatross in this light in a well-known essay of 1946, claiming that 'The albatross is the symbol of Coleridge's creative imagination, his eagle.'[49] In fact, Coleridge's recurrent choice of a bird to represent his poetic persona is the ostrich rather than the eagle. In an extended image in marginalia of 1833, for example, the aspiring poet contrasts himself with Jove-as-eagle – the 'Sovereign of the air ... who makest the mountain pinnacle thy perch and halting-place,' – by showing how, 'Like the ostrich, I cannot fly, yet I have wings that give me the feeling of flight', and by claiming that 'the linnet, the thrush, the swallow, are my brethren'.[50] Similarly in Chapter II of *Biographia Literaria* (1817) Coleridge laments:

> I have laid too many eggs in the hot sands of the wilderness of the world, with ostrich carelessness and ostrich oblivion. The greater part indeed have been trod under foot, and are forgotten; but yet no small number have crept forth into life, some to furnish feathers for the caps of others, and still more to plume the shafts in the quivers of my enemies, of them that unprovoked have lain in wait against my soul.
> *(Major Works,* ed by Jackson, p. 181)

The images here are largely drawn from classical iconography and, although Coleridge may have seen ostriches in Windsor Great Park where they were kept at this time by the Duke of Cumberland, the tour-de-force of extended imagery in this passage relies not on features observed from life in real birds but on the ostrich's fabled fecklessness to convey its meaning. Similarly, although one of his best-known natural images – that of the water-insect in Chapter VII of *Biographia Literaria* – appears initially to illustrate exactly the principle that everything (even a tiny insect) 'has a life of its own', this motif in fact turns out, as Coleridge himself points out, to be an 'emblem': an instance of the writer describing an 'interesting appearance in nature' precisely in order to 'connect ... it with dim analogies with the moral world':

---

[48] *The Notebooks of Samuel Taylor Coleridge* ed by Kathleen Coburn, 4 vols (London: Routledge & Kegan Paul, 1957–90) I. 1590.

[49] See also *Coleridge Notebooks* III. 3314, entry for 16 May 1808: 'O that sweet bird! where is it? ... It is in Prison – all its instincts ungratified – yet it feels the influence of spring & calls with unceasing melody the Loves, that dwell in Fields and Greenwood bowers – unconscious perhaps that it calls in vain. – O are they the songs of a happy enduring Day-dream? has the Bird Hope? Or does it abandon itself to the Joy of its Frame – a living Harp of Eolus? – O that I could do so! –'

[50] Marginalia written in a copy of J. H. Heinrichs' 1821 edition of the Book of Revelation, in *Samuel Taylor Coleridge: A Critical Edition of the Major Works,* ed by H. J. Jackson (Oxford: Oxford University Press, 1985) p. 567.

> Most of my readers will have observed a small water-insect on the
> surface of rivulets, which throws a cinque-spotted shadow fringed
> with prismatic colours on the sunny bottom of the brook; and will
> have noticed, how the little animal *wins* its way up against the stream,
> by alternate pulses of active and passive motion, now resisting the
> current, and now yielding to it in order to gather strength and a
> momentary fulcrum for a further propulsion. This is no unapt emblem
> of the mind's self-experience in the act of thinking. There are
> evidently two powers at work, which relative to each other are active
> and passive. *(Major Works* ed by Jackson, p. 222)

Coleridge is characteristic here in taking the motions of an animal to provide a
metaphor for the *human* mind at work and, as Trevor Levere remarks,

> When Coleridge went to the ant, it was not only with a naturalist's
> eye, but also with a view to extending his comprehension of the role
> of powers in mind and nature, the correspondences among them, the
> teleology directing them, and the language embracing and
> incorporating them.[51]

Coleridge's characteristic use of animals may, then, have more in common
than he would have liked to acknowledge with the approach taken by the
children's authors whom he and his circle mocked and deplored. Like them,
he primarily features animals as objects in human culture and specifically as
emblems or fables to illustrate human concerns or behaviour. A famous
passage from the 1805 *Notebook* demonstrates how pervasive this approach is
in his observation of natural objects in general: they are framed and altered by
a human viewpoint and the action of the mind, as the moon is here by the
window and the dew on the glass:

> In looking at objects of Nature while I am thinking, as at yonder moon
> dim-glimmering thro' the dewy window-pane, I seem rather to be
> seeking, as it were *asking*, a symbolical language for something
> within me that already and for ever exists, than observing anything
> new. *(Notebooks,* II. 2546)[52]

This approach can be found, moreover, among adherents of all the factions of
the educational debate with which this chapter opened, however these may be
construed in ideological terms. In the 'imagination/instruction' dichotomy, the
deployment of animals as emblems and fables is characteristic no less of the

---

[51] Trevor Levere, *Poetry Realized in Nature: Samuel Taylor Coleridge and Early
Nineteenth-Century Science* (Cambridge: Cambridge University Press, 1981) p. 213.

[52] Coleridge's account of his encounter with a lizard on Malta just before this (July 1804,
Notebooks, II. no. 2144) is also 'framed' and anthropomorphized by the human observer's
imagination: 'Lizard green with bright gold spots all over – firmness of its stand-like feet,
where the Life of the threddy Toes makes them seem & be so firm, so solid – yet so very, very
supple / one pretty fellow, whom I had fascinated by stopping & gazing at him as he lay in a
(thick) network of Sun & Shade ... then turned his Head to me, depressed it, & looked up half-
watching; half-imploring, at length taking advantage of a brisk breeze that made all the
Network dance & Toss, & darted off as if an Angel of nature had spoken in the breeze – Off!
I'll take care, he shall not hurt you.'

work of Coleridge and other 'imaginative' authors than it is of the books of 'instructive' writers for children; and followers of Locke and of Rousseau are alike in using this technique to pursue their arguments. So while Coleridge deploys the water-insect as 'no unapt emblem of the mind's self-experience in the act of thinking', Trimmer recounts fables of bird-life to train children in obedience and the duties of society, and Day tells stories of country and city dogs to show how urban civilization corrupts and weakens the (human) spirit.[53]

In Coleridge's case, one reason for this stylistic subordination of animals to humankind lies in his theological interests and religious convictions which, like Trimmer's, drew on traditional Christian doctrine about the immortality of the human soul and the unbridgeable gap between human and animal nature. His arguments on this subject are subtle and eclectic, producing a clever synthesis of Renaissance theriophilic ideas such as Montaigne's and mediaeval thinking associated with Thomas Aquinas:

> But if it should be asked, why this resurrection, or re-creation is confined to the human animal, the answer must be – that more than this has not been revealed. And so far all Christians will join assent. But some have added, and in my own opinion much to their credit, that they hope, it may be the case with the Brutes, likewise, as they see no sufficient reason to the contrary. And truly, upon *their* scheme, I agree with them. For if a Man be no other or nobler Creature *essentially,* than he is represented in their system [ie the Socinian] the meanest reptile, that maps out its path on the earth in lines of slime, must be of equal worth and respectability, not only in the sight of the Holy One, but by a strange contradiction even before Man's own reason. For remove all the sources of Esteem and Love founded on Esteem, and whatever else pre-supposes a Will and, therein, a possible transcendence to the material world: Mankind, as far as my experience has extended ... are *on the whole* distinguished from the other Beasts incomparably more to their *disadvantage,* by Lying, Treachery, Ingratitude, Massacre, Thirst of Blood, and by Sensualities which both in sort and degree it would be libelling their Brother-Beasts to call *bestial,* than to their advantage by a greater extent of Intellect.[54]

Anya Taylor believes that such arguments are the result of Coleridge's anxiety to resist the earliest intimations of the evolutionary debate, which culminated some fifty years later in *The Origin of Species:* '[H]e works to define the human being in contradistinction to the beast, in a manner that indicates a prophetic and Victorian distress,' she says.[55] Coleridge himself connects his fears with an earlier member of the Darwin family – Erasmus, the grandfather

[53] Richard E. Brantley, in *Wordsworth's 'Natural Methodism'* (New Haven: Yale University Press, 1975) describes Wordsworth's poems as 'seemingly pure descriptions but actually resonant allegorizings of birds, blossoms, meadows, mountains and the sea ... "naturalized emblemology"' (pp. 160, 166).

[54] 'A Lay Sermon Addressed to the Higher and Middle Classes' (1817): *Coleridge, Lay Sermons,* ed by R. J. White, *The Collected Works of Samuel Taylor Coleridge,* Bollingen Series (London: Routledge & Kegan Paul, 1972) p. 183.

[55] Anya Taylor, *Coleridge's Defense of the Human* (Columbus: Ohio State University Press, 1986) p. 40.

of Charles – when he admonishes Wordsworth for not rebutting in *The Excursion* the theory that the brains of orang-utans and of men are similar:

> I understood that you would take the Human race in the concrete, and have exploded the absurd notion of Pope's Essay on Man, Darwin, and all the countless Believers – even (strange to say) among Xtians of Man's having progressed from an Ouran Outang state – so contrary to all History, to all Religion, nay to all Possibility. *(Collected Letters,* IV. 574–5)[56]

For Coleridge, as for most pre-Darwinian religious believers, humankind is eternally and incontrovertibly at the centre of the universe, and metaphors and analogies which draw upon the natural world to explain humanity and to show the 'oneness' between human beings and other living things most strongly serve to demonstrate the benevolence of the Deity in creating a world where all things are fitted to human use. This world-picture gains strength from the unity-in-diversity of the universe as perceived by its believers and it is not in itself a stance which disadvantages animals, since their very similarity with humankind demonstrates the way in which they should be benevolently treated. Where this circle is broken, in the Romantic period, it is usually because of religious scepticism or doubt about the existence of a benevolent deity, and this is the situation which is explored in Chapter Six. The contrast between Lockean and Rousseauian concepts about animals, and their deployment in educational theory, is also highly relevant to the parliamentary arguments about improving the treatment of animals and educating the human perpetrators of cruelty, and this forms the subject of the following chapter.

---

[56] The theory was expressed by Edward Tyson in his influential book *Orang-Outang, sive Homo Sylvestris; or, The Anatomy of a Pigmie Compared to that of a Monkey, an Ape, and a Man* (1699). Wordsworth succeeds in avoiding or evading this issue in his work in general, and I demonstrate in Chapter Five how he classifies animals ambiguously: neither as part of human culture, but nor as part of inspirational Nature.

Chapter 3

Political Animals:
Bull-fighting, Bull-baiting
and *Childe Harold* I

On 15 May 1809 Thomas Lord Erskine, legal defender of Horne Tooke and others in the Treason Trials of 1794 and veteran of the slavery abolition campaign, introduced a bill in the House of Lords which he described as both 'a new aera in the history of the world' and 'a subject very dear to my heart'.[1] The measure he hoped to see enacted was a bill to prevent 'wanton and malicious cruelty' to domestic animals in general – the first of its kind ever to be debated in any Western legislature. Erskine was careful, however, to avoid any specific mention of bulls in his draft, since attempts to prevent bull-baiting, introduced by William Wilberforce and his associates in 1800 and 1802, had been roundly defeated then in the House of Commons by the almost single-handed efforts of William Windham, MP for Norwich and erstwhile Secretary at War.

Erskine's bill too was destined to fail, as were numerous subsequent attempts to introduce this type of legislation in the next decade, until a bill of 1822 (introduced by Richard Martin MP in the Commons and piloted through the Lords by Erskine) succeeded in providing protection for 'Horses, Mares, Geldings, Mules, Donkeys, Cows, Heifers, Bull Calves, Oxen, Sheep, and other Livestock'.[2] A reporter from *The Times* recorded the ridicule encountered by one such attempt in 1821, when an MP suggested that 'asses should also be protected from the cruelty to which they were exposed':

> We could not catch the particulars of his remarks owing to the noise and laughter which prevailed. ... Mr Monck [said, if this bill] should pass, he should not be surprised to find some other member proposing a bill for the protection of dogs (a member here said 'and cats'); and he had no doubt that if there were any members of that house attached to cats, that they might also hear a bill respecting them. (Hear hear and laughter.)[3]

---

[1] *The Parliamentary Debates from the Year 1803 to the Present Time*, vol. XIV (London: Hansard, 1812) pp. 571 and 553. Hereafter referred to in the text as *Debates*, XIV. The debates were recorded in a mixture of direct and reported speech.

[2] *The Parliamentary Debates from the Year 1803 to the Present Time*, n.s., vol. VII, (London: Hansard, 1825) p. 758. This legislation still excluded wild and pet animals from protection: reference to the latter having been deleted in the House of Commons. Legislation banning baiting was not passed until 1835.

[3] *The Times*, Saturday 2 June 1821, p. 3. Dogs were not given legal protection until 1839. The cruel commercial exploitation of seals (as evoked by Keats, in *Isabella* stanza xv: 'for them in death / The seal on the cold ice with piteous bark / Lay full of darts') has only recently been addressed. *Keats, Poetical Works*, ed by H. W. Garrod (London: Oxford University Press, 1959, repr. 1967).

Erskine's 1809 speech to the House of Lords was a more decorous affair, and was given a respectful hearing by an audience of over one hundred peers, including two royal dukes and several bishops and archbishops.[4] The address was widely reported in the press and subsequently published as a pamphlet, but among those who were privileged to hear it 'live', as it was first delivered, was the twenty-one-year-old Lord Byron, who had first taken his seat in the House of Lords two months earlier and was at this time actively considering a parliamentary career.[5] Two months later Byron attended the bullfight near Cadiz which is the subject of stanzas 68 to 80 of Canto I of *Childe Harold's Pilgrimage,* written in Thebes at the end of that year.

One of the questions to be discussed in this chapter is whether the verse Byron wrote in December 1809 about the scene he saw in July reflects the political debate he heard in May. The chapter also examines the parliamentary debates of 1800 and 1802 about bull-baiting, as well as the 1809 debate about cruelty to animals in general: reading these and Byron's verse about bullfighting in the context of wider Burkean and anti-Burkean stances in English political and cultural discourse in the early nineteenth century. It shows how frequently these literary and political materials interchange and borrow each other's features and methods, and how they both make use of and participate in cultural issues on a much wider plane than that with which they are ostensibly concerned. Such a reading demonstrates how Byron's presentation of the Spanish bullfight reflects some very English arguments which relate not only to the rights and treatment of animals but also to their role in the formation of what William Windham called 'the old English character'.[6] Drawing on the 1800, 1802 and 1809 debates, it locates Byron's parodically chivalric treatment of the bullfight, and his characterization of the animals involved, alongside the complex antitheses between tradition and innovation, Englishness and foreignness, Gothicism and classicism, romantic chivalry and practical politics which, in the decades following Burke's *Reflections on the Revolution in France* (1790), were the site of powerful arguments in British political rhetoric.

This series of 'animal cruelty' debates has been read as part of the social history of the development of humanitarianism and in the light of Lord Macaulay's famous gibe (in his *History* of 1848) that 'The Puritan hated bear-baiting not because of the pain it gave to the bear, but because it gave pleasure to the spectators'.[7] The supporters of the 1800 and 1802 bills did indeed have a programme of human social amelioration in mind, in that they believed cruel sports such as bull-baiting morally depraved the participants and, as Wilberforce

---

[4]   Erskine claimed the following year that there had been a hundred and ten peers present, although the *Journal of the House of Lords* XLVII. 253 records the names of only seventy-seven.

[5]   Byron attended the House of Lords forty-eight times in all, and on fifteen occasions he was appointed to a select committee to consider various bills: a way of serving a parliamentary apprenticeship. He delivered three speeches in the Lords, including his maiden speech in defence of the frame-breakers in February 1812. See *LBCMP*, pp. 278–9.

[6]   *The Parliamentary History of England,* vol. XXXVI (London: Hansard, 1820) p. 833. Hereafter referred to in the text as *History,* XXXVI.

[7]   See E. S. Turner, p. 39, and Macaulay, *The History of England from the Accession of James II,* 3 vols (London: J. M. Dent, 1906, repr. 1934) I. 129. Macaulay may have drawn his argument from Windham's speech in the 1802 debate.

claimed in the 1802 debate (in phrases which linked the human perpetrators of cruelty in terms of value with the animals they tormented), 'degraded human nature to a level with the brutes'. Sir William Pulteney's introduction of the 1800 bill made it evident that he was seeking to protect the human beings, rather than the animals involved – including the masters of the 'idle and disorderly persons' who were attending bull-baits instead of getting on with their work. Bull-baiting, he declared,

> was cruel and inhuman; it drew together idle and disorderly persons; it drew also from their occupations many who ought to be earning a subsistence for themselves and their families; it created many disorderly and mischievous proceedings, and furnished examples of profligacy and cruelty.[8]

Sir Richard Hill was led to second the measure 'from a love of decency and decorum, and out of humanity to the common people' *(History,* XXXV. 203).

William Windham

The apparent political naiveté of some of the supporters of these bills was matched by considerable rhetorical sophistication on the part of their major opponent, William Windham. Among the fulsome tributes after Windham's death in 1810 was that of an un-named parliamentarian who said of his oratory: 'If it was not the most commanding that that house has ever seen, it was the most insinuating'.[9] In similar vein, Byron praised Windham in 1813 as

> the first in one department of oratory and talent, whose only fault was his refinement beyond the intellect of half his hearers. ... I, who have heard him, cannot regret any thing but that I shall never hear him again ... he is gone, and Time 'shall not look upon his like again.' *(Journal,* 24 November 1813, *BLJ* III. 219)

By 1821, however, Byron's opinion had changed: 'Windham I did not admire though all the world did – it seemed sad sophistry,' he recorded in his *Detached Thoughts* for October of that year *(BLJ* IX. 14). John Lawrence (a supporter of anti-cruelty measures) gave a clearer and harsher picture of Windham's style in *The Sportsman's Repository* (1820) when he commented:

> That man had in an eminent degree the gift of the gab; and at the same time, the pre-eminent art of confounding every subject beyond all possibility of its being developed and comprehended

---

[8] *The Parliamentary History of England,* vol. XXXV (London: Hansard, 1819) p. 202. Hereafter referred to as *History* XXXV. Wilberforce complained of Pulteney's performance that he had 'argued it like a parish officer, and never once mentioned the cruelty': letter to Hannah More, 25 April, 1800, quoted in Robert I. and Samuel Wilberforce, *The Life of William Wilberforce,* 2 vols (London: John Murray, 1838) II. 366.

[9] Thomas Amyot, *Speeches in Parliament of the Right Honourable William Windham: To which is Prefixed some Account of his Life* (London: Longman, 1812) p. 134.

either by himself or others. He was the very Hierophant of
confusion and his mind the chosen Tabernacle of that goddess: he
had in truth been so much in the habit of shaking up right and
wrong in the bag together that he had long lost the faculty of
distinguishing one from the other. (E. S. Turner, p. 111)[10]

Windham's opposition to the bull-baiting and animal cruelty bills formed a major
part of his reputation, and his epitaph picks out what was the most prevalent
feature of his performance in this field in order to characterize his achievement in
general: 'He was, above all things, anxious to preserve, untainted, the National
Character, and even those National Manners which long habit had associated
with that character' (quoted by E. S. Turner, p. 110).

Windham was a friend and political pupil of Burke, and following Burke he
took alarm at the possible effects on British political life of the French
Revolution and in the early 1790s became one of the most ardent supporters of
the government's repressive legislation.[11] His politics, like Burke's, were
complicated by changing allegiances and party affiliations, and he won the
nickname 'Weathercock Windham' because, having been leader of the
opposition to Castlereagh and to the French War, he changed face during the
Fox-Grenville ministry of 'All the Talents' by becoming a supporter of the
War.[12] His opposition to the animal protection bills shows not only the
hallmarks of Burke's style, with the rhetorical flourishes, the measured periods,
the effective use of anecdote and example, but also a thorough mastery of
Burke's ideas on revolution and counter-revolution and the application of them
to a completely new area of discourse. Drawing on Burke's distinctions
between the old and the new, organic conservatism and rational change,
Windham succeeded in introducing fundamental issues of class, gender, history,
national character and feeling into the debates on animal protection. It was,
Windham said, 'absurd to legislate against the genius and spirit of the country.
The putting of a stop to bull-baiting was legislating against the genius and spirit
of almost every country and every age' (*History*, XXXV. 212). The
exploitation of this powerful ideology in a field which seemed totally unrelated
to it took the supporters of these bills by surprise and left them floundering.
Windham converted what they had seen as a non-party issue into an intensely
political topic, by questioning the patriotism, sense and even the good faith of
those who introduced it. During the earlier debates, only the histrionic abilities
of Richard Brinsley Sheridan, in 1802, came close to countering Windham's
powerful appeal to tradition, and exposing its sophistries by passages of strong
emotional appeal which matched those of his opponent.

Windham's arguments in the two sets of debates (on bull-baiting in 1800 and
1802, and on cruelty in general in 1809) were not identical. In the earlier
debates, in his capacity as Secretary at War, he made use of the war situation to
belittle the subject of the bill:

---

[10] Lawrence was the author of *A Philosophical and Practical Treatise on Horses: And on
the Moral Duties of Man towards the Brute Creation,* 2 vols (London: 1796–98).

[11] See *Dictionary of National Biography* ed by Leslie Stephen and Sidney Lee, 63 vols and
supplement (London: Smith, Elder, 1885–1901) LXII. 173.

[12] Byron described him in January 1809 as 'Mr Windham with his Coat *twice* turned' *(BLJ*
I. 186).

Really, Sir, in turning from the great interests of this country and of Europe, to discuss with equal solemnity such measures as that which is now before us, the House seems to me to resemble Mr Smirk, the auctioneer in the play, who could hold forth just as eloquently upon a ribbon as upon a Raphael. *(History, XXXV. 204)*

Windham compared the supporters of the bills to 'Jacobins and Methodists'. By the Methodists, he claimed,

every rural amusement was condemned with a rigour only to be equalled by the severity of the puritanical decisions. They were described as part of the lewd sports and Antichristian pastimes which in times of puritanism had been totally proscribed. Every thing joyous was to be prohibited, to prepare the people for the reception of their fanatical doctrines. By the Jacobins on the other hand, it was an object of important consideration to give to the disposition of the lower orders a character of greater seriousness and gravity, as the means of facilitating their tenets; and to aid this design, it was necessary to discourage the practice of what were termed idle sports and useless amusements. This was a design which he should ever think it his duty strenuously to oppose. *(History, XXXVI. 834)*

Presenting the attempts to ban bull-baiting as 'directed to the destruction of the old English character' (p. 834), Windham maintained that 'The habits long established among the people were the best fitted to resist the schemes of innovation; and it was among the labouring and illiterate part of the people that Jacobinical doctrines had made the smallest progress' (p. 834).

Neatly bringing together the two strands of his argument in order to illustrate that conservatism in this matter best accorded with the efficient prosecution of the war, he claimed that 'the counties of Lancashire and Staffordshire, where the practice [of bull-baiting] principally prevailed ... were known to produce the best soldiers for the army' *(History, XXXVI. 840)* – presumably because the locals plentifully consumed the courage-promoting flesh of the bulls they baited.[13] This particular argument of his became notorious, and it was the subject of a cutting reference by Hobhouse in a footnote to Byron's note on line 1267 of *Childe Harold* IV (1818 – part of the 'Coliseum' passage).[14] Byron had

---

[13] Dr William Lambe, whose research on vegetarianism was a source for Shelley's *Vindication of a Natural Diet,* claimed the opposite causes to account for the high physical quality of the inhabitants of this region: 'The peasantry of Lancashire and Cheshire, who live principally on potatoes and buttermilk, are celebrated as the handsomest race in England': *Additional Reports on the Effects of a Peculiar Regimen in Cases of Cancer, Scrofula, Consumption, Asthma, and other Chronic Diseases* (London: Mawman, 1815) p. 220.

[14] See *BCPW* II. 338, quoting from Hobhouse's *Historical Illustrations to the Fourth Canto of Childe Harold* (London: Murray, 1818). Hobhouse accompanied Byron on almost all the sightseeing visits in Rome which gave rise to this Canto and, as McGann points out, the notes 'are important not only in themselves, but also for the information they provide about the books which B. and H. were reading at the time' *(BCPW* II. 317). When Hobhouse invokes Windham's speeches as commentary on *Childe Harold* IV it is therefore fair to assume that he and Byron had discussed the Coliseum and its history with reference to these.

written with disdain of the commentator Lipsius's idea that 'the loss of courage, and the evident state of degeneracy of mankind' at the end of the Roman Empire could 'be nearly connected with the abolition of ... bloody spectacles' in the amphitheatre: implying that conversely the frequent gladiatorial combats might actually have contributed to the degeneracy of the spectators *(BCPW*, II. 258). Hobhouse described Lipsius's commentary ironically as 'the prototype of Mr Windham's panegyric on bull-baiting' *(BCPW*, II. 338), comparing Windham's idea that participation in bull-baiting had given British men their fighting spirit with the theory that the 'torrents of blood' shed in the circus had contributed to ancient Roman martial greatness.

Windham turned the concern of the bills' proposers for the morals of the poor on its head by maintaining that bull-baiting was actually *good* for the lower orders. He pointed out that it was one of the few amusements from which they were not excluded by their poverty. Even 'To dance at all out of season was to draw on their heads the rigour of unrelenting justice,' and 'it was known that an organ did not sound more harshly in the ears of a puritan than did the notes of a fiddle within those of a magistrate, when he himself was not to be of the party' *(History,* XXXVI. 837). Burke's *Reflections,* which had aimed to defend the threefold order of the monarchy, the nobility, and the (French) church, indicated Burke's own Roman Catholic connections by including a sympathetic treatment of some aspects of Catholicism. Claiming that England still cherished the 'Gothic and monkish education', Burke had maintained that England was still 'so tenacious ... of the old ecclesiastical modes and fashions of institution, that very little alteration has been made in them since the fourteenth or fifteenth century' and that 'Superstition is the religion of feeble minds; and they must be tolerated in an intermixture of it, in some trifling or some enthusiastic shape or other, else you will deprive weak minds of a resource found necessary by the strongest'.[15] Similarly, Windham unfavourably contrasted 'puritanism' in England, which put a stop to rural festivities such as 'hops', with the state of the common people in Roman Catholic countries such as 'the south of France and in Spain, [where] at the end of the day's labour, and in the cool of the evening's shade, the poor dance in mirthful festivity on the green, to the sound of the guitar' *(History,* XXXV. 205). In such countries

> they enjoy many more amusements and a much longer time for relaxation than the poor in this country, who may say with justice, 'Why interfere with the few sports that we have while you leave to yourselves and the rich so great a variety?' (p. 204)

---

[15] Edmund Burke, *Reflections on the Revolution in France,* ed by A. J. Grieve (London: Everyman, Dent, 1910, repr. 1967) pp. 96–7 and 155. Burke's invocation of 'the Gothic' was in deliberate contrast to the way in which the revolutionary debate in both France and England was widely characterized by use of examples from Roman history and Roman writings. Jane Worthington, in *Wordsworth's Reading of Roman Prose* (Newhaven: Yale University Press, 1946) p. 3, has shown how Roman writers were cited in debates and newspapers in the National Assembly, Legislative Assembly, and the National Convention, and how 'John Thelwall offers the most striking example of an Englishman using ancient history after the French manner. In 1796 when the convention Bill prevented his direct attacks upon the British Government, he immediately resorted to lectures on Roman history. In these lectures ... he

Windham pointed out how 'in the time of Queen Elizabeth, that which was now despised and reprobated as the amusement only of the lowest of the people, was an amusement courted by all ranks' *(History,* XXXVI, p. 836); and in an argument highly reminiscent of Burke's defence of the monarchy and the old order in general in the *Reflections,* he claimed (p. 839) that

> the antiquity of the thing was deserving of respect, for respect for antiquity was the best preservation of the church and state – it was by connecting the past with the present, and the present with the future, that genuine patriotism was produced and preserved.

Windham's argument about the ancient respectability of bull-baiting was well founded: as E. S. Turner has pointed out (p. 34), in the seventeenth century,

> the flesh of the bull was thought to be indigestible and unwholesome if the animal was killed without being baited and butchers were liable to prosecution if the brutal preliminaries were omitted. Numerous butchers were still being fined for this in Stuart times.[16]

In the previous century, the view of Queen Elizabeth and her ministers was that bull-baiting was a more fit national pastime than the theatre, which put new and irreverent ideas into people's heads, and a Lord Mayor of London commented that 'in divers places, the players do use to recite their plays, to the great hurt and destruction of the game of bull-baiting and such-like pastimes which are maintained for Her Majesty's pleasure'. The Queen herself asked for an order to restrain the production of plays on Thursdays because that was bull-baiting day (E. S. Turner, p. 35).

Windham's appeal to history also took in references to Gratian's and Claudian's descriptions of the fame of the British bull-dog in ancient times, and his literary allusions ranged from quoting Goldsmith, Gray and other poets, to an evocation of

> a beautiful passage of Sterne, in which he described the lower orders at the close of the day, when labour was finished, when families met together to join in social pleasures, when the old encouraged the sports of the young and rejoiced in the amusements of their children. *(History,* XXXVI. 837)

Unlike Burke's, however, Windham's attitude to women was not especially partial or chivalrous. Burke's idea of the beautiful had been intimately connected with smallness and weakness, both in women ('so far is perfection, considered as such, from being the cause of beauty; that this quality, where it is highest in the female sex, almost always carries with it an idea of weakness and

---

continued to expound republican principles, merely shifting the point of application from modern England to ancient Rome' (p. 10).

   [16] Edwards (in his section on 'Canis Pugnax, The Bull-dog') comments 'The idea that the flesh of the bull is rendered more tender from being baited, was perhaps another cause for the frequency of this sport; and I believe there is still an act of parliament unrepealed, forbidding, under pains and penalties, the selling of bull meat unless it had been baited.'

imperfection' *[Sublime and Beautiful,* p. 110]), and in animals ('In the animal creation, it is the small we are inclined to be fond of; little birds, and some of the smaller kinds of beasts' [p. 113]). In Windham's vision, however, the microscopic view of a woman's face takes on nightmarish, Swiftian dimensions: 'if even the fairest complexion were contemplated through a microscope, deformities would appear, and hairs unobservable to the naked eye would present themselves as the bristles on the back of a boar' *(History,* XXXVI. 833).

Burke's famous evocation of Marie Antoinette had envisioned her as 'glittering like the morning-star, full of life, and splendour, and joy'; had imagined how 'ten thousand swords must have leaped from their scabbards to avenge even a look that threatened her with insult,' and had lamented that

> the age of chivalry is gone. ... Never, never more shall we behold that generous loyalty to rank and sex, that proud submission, that dignified obedience, that subordination of the heart, which kept alive, even in servitude itself, the spirit of an exalted freedom. (Burke, *Reflections,* p. 73)

Windham in the 1802 debate, in contrast, subjected women to a brisk mimicry which satirized the 'species of philosophy' dictated to gentlemen by their wives:

> 'My dear, do you know, that after you went out with your dogs this morning, I walked into the village, and was shocked to see a set of wretches at a bull-bait, tormenting the poor animal. I wish, dear, you would speak to our member, and request him to bring a bill into parliament to prevent that horrid practice.' *(History,* XXXVI, p. 840)

In the 1809 debate Windham went further, and used one of the lively social vignettes which helped to give his oratory its popular appeal specifically to address the idea that, as Sir Richard Hill had claimed in 1802, 'the amiable sex' were particular advocates of humanity to animals. The 'fashionable female circles' who appeared to Windham to have been 'very diligently canvassed' by the supporters of the bill, were on the contrary he claimed, some of the prime causes of inhumanity to animals, and the perpetrators of what he particularly criticized: the way in which the bill would fall heavily upon the poorer classes by threatening them with harsh penalties for animal cruelty, while leaving the rich untouched. These ladies alleged, he said, that they had been 'continually shocked' by cases of coachmen whipping their horses in public places:

> But apply to any of these ladies, and satisfy them, after much difficulty, that their coachman was most active and most in the wrong, in the struggle, which caused so much disturbance at the last Opera, and the answer probably would be, 'Oh! to be sure; it is very shocking; but then John is so clever in a crowd! the other night at Lady Such-a-one's, when all the world was perishing in the passage, waiting for their carriages, ours was up in an instant, and we were at Mrs Such-a-one's half an hour before anyone else. We should not know what to do, if we were to part with him.' *(Debates,* XIV. 1037)

'Was it a coachman here who most deserved punishment?' Windham pointedly asked (p. 1037).

Sheridan seems to have been alone among Windham's opponents, in the earlier debates, in grasping how cleverly the Norwich MP had used what were actually deeply conservative arguments to make himself appear as the champion of the rights of the poor. Sheridan claimed that Windham's position was only 'the mask of friendship for the people': if Windham wished, under cover of this mask, 'to make them servile, he would teach them to be cruel. If he wished to induce them to submit to a system of government by barracks and bastiles, he would encourage bull-baiting' *(History,* XXXVI. 853). Sheridan attempted to counter Windham's arguments by returning to the ostensible subjects of discussion: the suffering of the baited animals and the ill-effects of cruelty on the morals of the poor. Using a Rousseauian idea, he pointed out that

> many of the most striking lessons to man were to be learned from animals, but it was from animals in their natural state and exhibiting their natural qualities. ... It was not by using craft to make an animal the enemy of another (not so by nature), that instruction was to be obtained. *(History,* XXXVI. 852)

He painted an effective word-picture of the savagery and pathos involved in the sport:

> What sort of moral lesson, for instance, was it to the children of the farmer, who brings his aged bull-bitch, many years the faithful sentinel of his house and farmyard, surrounded by her pups, to prove at the bull-ring the staunchness of her breed? He brings her forward, sets her at the infuriated animal; she seizes him by the lip, and pins him to the ground. But what is the reward from his owner, amid the applauses of the mob, to his favourite animal? He calls for a hedging bill, and to prove her breed, hews her to pieces without her quitting her grip, while he sells her puppies at five guineas apiece! Another enters his dog at the animal; his leg is broken in the attack. His owner lays a wager that he shall pin the bull near the lip. He calls the dog, cuts off his leg, then sets him at the bull, which he pins; and having thus won his wager, he is whistled back to his grateful master who, while he licks his hands, generously cuts his throat. *(History,* XXXVI. 852)

Sheridan also tried humour, comparing the tenacious bull-dogs ('sullen, stubborn, and treacherous') to political placemen who 'when they had once laid hold of any thing they never let go their hold' *(History,* XXXV, p. 213). The joke raised a laugh, but his efforts of this and all other kinds were unavailing, and not helped by the lacklustre performance of Wilberforce himself in the 1802 debate. Windham succeeded in defeating both measures, albeit by quite narrow margins: by 43 votes to 41 in the first debate, and by 64 to 51 in the second.

Lord Erskine

In the 1809 debates, Windham had a more astute and, above all, much more committed, opponent in Lord Erskine: a former colleague from the coalition government of 1806–07, where Windham had been Colonial Secretary to Erskine's Lord Chancellor.[17] Erskine's weak point was his long-windedness: Byron, who later came to know him personally, several times commented on his notorious egoism in company, while continuing to respect him sincerely for 'a life so dear to Fame and Freedom'; and in the official records Erskine's speeches are interspersed with signs of impatience from his audience.[18]

Erskine's strength lay in his love of animals, amounting to a cultivated eccentricity. Lloyd P. Strycker in his biography includes several anecdotes about this fondness of Erskine's, including these related by Sir Samuel Romilly (one of the 1809 bill's defenders in the Commons):

> He has always expressed and felt a great sympathy for animals. He has talked for years of a bill he was to bring into Parliament to prevent cruelty towards them. He has always had several favourite animals to whom he has been much attached, and of whom all his acquaintance have a number of anecdotes to relate – a favourite dog which he used to bring, when he was at the Bar, to all his consultations; another favourite dog, which at the time when he was Lord Chancellor, he himself rescued from the street from some boys who were about to kill him, under pretense of its being mad; a favourite goose which followed him wherever he walked about his grounds; a favourite mackaw, and other dumb favourites without number.[19]

One of Erskine's most eccentric animal moments was at a dinner held at his house in January 1808:

> He told us that he had two favourite leeches [Romilly recorded]. He had been blooded by them last autumn ... they had saved his life and he had brought them with him to town; had ever since kept them in a glass; had himself every day given them fresh water; and had formed a friendship with them. He said he was sure they both

---

[17] Erskine and Windham never faced each other directly in the animal cruelty debates, of course, since one was a member of the House of Lords and the other of the House of Commons.

[18] Byron wrote in his 'Detached Thoughts' (1821–22) *BLJ* IX. 44: 'In 1812, at Middleton (Lord Jersey's) amongst a goodly company – of Lords – Ladies – & wits – etc ... Erskine – was there – good – but intolerable – he jested – he talked – he did everything admirably but then he would be applauded for the same thing twice over – he would read his own verses – his own paragraph – and tell his own story – again and again – and then "the trial by Jury!!!" – I almost wished it abolished, for I sate next to him at dinner – And as I had read his published speeches – there was no occasion to repeat them to me.' For Byron's political respect for Erskine, see a letter of 10 November 1823 to Erskine: *BLJ* XI. 59.

[19] Lloyd Paul Stryker, *For the Defence: Thomas Erskine, One of the Most Enlightened Men of his Times, 1750–1823* (London and New York: Staples Press, 1949) p. 486.

knew him, and were grateful to him. He had given them different names, Home and Cline (the names of two celebrated surgeons), their dispositions being quite different. After a good deal of conversation about them, he went himself and brought them out of his library, and placed them in a glass upon the table.
(Strycker, p. 486)

Strycker suggests that it was Byron's influence that led Erskine into trying his hand at verse, and quotes the following example (p. 98) about the pony which had carried Erskine early in his career:

Poor Jack, thy master's friend when he was poor,
Whose heart was faithful and whose step was sure.
Should prosperous life debauch my loving heart,
And whispering pride repel the patriot's part;
Should my feet falter at ambition's shrine,
And for mere lucre quit the path divine;
Then may I think of thee when I was poor,
Whose heart was faithful and whose step was sure.

Erskine's introductory statement for the 1809 bill was still predicated on the assumption that animals are 'subservient to human purposes' *(Debates,* XIV. 555) but he nevertheless placed animals on quite a different footing from that of Wilberforce and his associates. His measure was based on the need for 'attending to the feeling of the animal itself', as he later put it, 'and preventing cruelty from a consideration of its suffering'.[20] He emphasized what animals have in common with human beings, rather than what keeps them separate:

For every animal which comes in contact with man, and whose powers, qualities, and instincts are obviously constructed for his use, nature has taken the same care to provide, and as carefully and bountifully as for man himself, organs and feelings for its own enjoyment and happiness. Almost every sense bestowed upon man is equally bestowed upon them; seeing, hearing, feeling, thinking; the sense of pain and pleasure; the passions of love and anger; sensibility to kindness, and pangs from unkindness and neglect, are inseparable characteristics of their natures as much as our own.
(p. 555)

This picks up, and advances upon, Bentham's Utilitarian position, quoted more fully in Chapter One:

The question is not, can they reason? nor, can they talk? But, can they suffer? Why should the law refuse its protection to any sensitive being? The time will come when humanity will extend its mantle over everything which breathes. We have begun by attending to the condition of slaves, we shall finish by softening that of all the animals which assist our labours or supply our wants.
(Bentham, p. 238)

---

[20] Erskine used these phrases in his attempt to reintroduce the bill in May 1810: see *Parliamentary Debates,* XVI (London: Hansard, 1812) p. 881.

Rather than using the Wilberforcean argument about the ill effects of cruelty upon its human perpetrators, Erskine firmly placed the animals' interests first:

> Animals are considered as property only: to destroy or to abuse them, from malice to the proprietor, or with an intention injurious to his interest in them, is criminal; but the animals themselves are without protection; the law regards them not substantively; they have no rights! *(Debates,* XIV. 554)

Conscious of the possibly threatening newness of this position, he made repeated attempts to reassure his hearers that it was in line with divine providence and with Man's dominion over animals: 'the justest and tenderest consideration of this benevolent system of nature is not only consistent with the fullest dominion of man over the lower world, but establishes and improves it' (p. 555); and he referred to several theological works and also to a well-known passage of Cowper's *The Task* in support of his position:

> If man's convenience, health,
> Or safety, interfere, his rights and claims
> Are paramount, and must extinguish their's.
> Else they are all – the meanest things that are –
> As free to live, and to enjoy that life,
> As God was free to form them at the first.
> (I. 581–6)

Erskine believed that his bill had been framed so as to avert clashes on various contentious areas, including bull-baiting:

> This Bill says not a word about bull-baiting. I only include the bull in my catalogue of protected animals. They, therefore, who support the practice, may still support it successfully, if they can convince court and jury ... that it does not fall within the description of wilful and wanton cruelty. *(Debates,* XIV. 560)

Instead, he concentrated his attention upon post-horses, which suffered grievously because innkeepers would

> devote an innocent animal to extreme misery, if not to death itself, by a manifest and outrageous excess of labour, rather than disoblige a mere traveller, engaged in no extraordinary business, lest in future he should go to the inn opposite. (p. 561)

Apart from his conviction, one of Erskine's most effective weapons was that used by Sheridan in the bull-baiting debates: graphic and harrowing descriptions of the actual cruelty practised, as when he read from a letter which described

> a very general practice of buying up horses not capable of being even further abused by any kind of labour. These horses, it appeared, were carried in great numbers to slaughter-houses, but not killed at once for their flesh and skins, but left without sustenance, and literally starved to death, that the market might be gradually fed. The poor animals, in the mean time, being reduced

to eat their own dung, and frequently gnawing one another's manes in the agonies of hunger. (p. 563)

In conclusion, Erskine stressed the newness of his measures: the bill, if passed 'will not only be an honour to the country, but an aera in the history of the world' (p. 571), and he pointed out that legislation was a powerful tool for changing perceptions on moral issues:

I must again impress upon your lordships' minds, the great, the incalculable effect of wise laws, when ably administered upon the feelings and morals of mankind. We may be said, my lords, to be in a manner new created by them ... From the moral sense of the parent re-animated, or rather in the branch created by the law, the next generation will feel, in the first dawn of their ideas, the august relation they stand in to the lower world, and the trust which their station in the universe imposes on them. (pp. 570–1)

It was on this aspect of the bill, when it reached the House of Commons in June, that Windham immediately pounced. His first and general objection to it was

that the object of it, however commendable, was not such as to become a fit subject of legislation. For this opinion he had at least a pretty strong voucher, in the universal practice of mankind down to the present moment. In no country had it ever been attempted by law to regulate the conduct of men towards brute animals, except in so far as such conduct operated to the prejudice of men. ... The novelty of the subject, not in its details or particular application, but in its general character, was a topic, not brought forward as an objection, by the opposers of the bill, but claimed and insisted upon by its authors. ... we ought to take care, to be cautious at least, how we began new eras of legislation. *(Debates,* XIV. 1029–30)

Windham's arguments were more subtle than they had been in 1800 and 1802, and he did not overtly attack the purpose of the bill:

Of the desirableness of the object, speaking abstractedly [sic], there could be no doubt. As far as wishes went, every man must wish, that the sufferings of all animated nature were less than they were. (p. 1030)

He even quoted a Utilitarian maxim:

Morality itself might perhaps be defined as, 'a desire rationally conducted to promote general happiness,' and consequently to diminish general pain. ... Let the duty be as strongly enforced, as far as precept and persuasion could go, and the feeling as largely indulged, as its most eager advocates could wish. (pp. 1030–1)

But this was only the preliminary skirmishing, and Windham soon moved on to the second major theme of his argument, which was that the bill would fall heavily upon or, as he put it, 'torment', the poorer classes, while leaving the rich untouched.[21] 'The bill, instead of being called a Bill for Preventing Cruelty to Animals, should be entitled, A Bill for harassing and oppressing certain classes among the lower orders of the people' (p. 1036). The question of field sports, which Erskine had said in Committee he hoped could be kept out of the debate, was not one which Windham was prepared to let go by:

> if with such a preamble on our statutes, and with acts passed in consequence to punish the lower classes for any cruelty inflicted upon animals, we continued to practise and to reserve in great measure to ourselves the sports of hunting, shooting and fishing, we must exhibit ourselves as the most hardened and unblushing hypocrites that ever shocked the feelings of mankind. (p. 1040)

And he added another of his vignettes mimicking the social discourse of upper-class life to clinch the point:

> What a pretty figure must we make in the world, if in one column of the newspapers we should read a string of instances of men condemned under the 'Cruelty Bill,' some to the county gaol to wait for trial at the assizes, some by summary process to the house of correction; and in another part, an article of 'Sporting Intelligence,' setting forth the exploits of my Lord Such-a-one's hounds, how the hounds threw off at such a cover; that bold Reynard went off in gallant stile, etc; and was not killed till a chace of ten hours; that of fifty horsemen who were out at the beginning not above five were in at the *death,* that three horses *died* in the field, and *several* it was thought would never *recover;* and that upon the whole it was the most glorious day's *sport,* ever remembered since the pack was set up! (p. 1040)

His argument was not, of course, that Parliament should proceed to abolish field sports, but that until the Parliamentary sporting gentlemen were prepared to cast the beam from their own eye, they were ill-placed to legislate out the mote which was in that of their poorer neighbours.

Windham concluded by referring again to the novelty of the proposed measure: 'There had grown up in the country, of late years, a habit of far too great facility in passing laws,' he claimed (p. 1026* [22]). He appealed to his fellow MPs not to 'run counter to the nature of things, by attempting what, as the authors of the bill themselves tell us, never yet was attempted' (p. 1027*);

---

[21] This aspect had, as Windham pointed out, been discussed in an anonymous critique in the *Edinburgh Review,* XIII. 26 (1809) pp. 339–41, about the proceedings of the Society for the Suppression of Vice, where the prevention of cruelty to animals was described as a specious attempt by the rich to exercise control over the amusements of the poor. 'Any cruelty may be practised to gorge the stomachs of the rich, – none to enliven the holidays of the poor. We venerate those feelings which really protect creatures susceptible of pain, and incapable of complaint. But heaven-born pity, nowadays, calls for the income tax, and the court guide; and ascertains the rank and fortune of the tormentor, before she weeps for the pain of the sufferer.' (p. 340).

[22] In this volume of the *Parliamentary Debates* (XIV) a sequence of pages numbered with a star (from 1025* to 1040*) is interpolated between the regularly-numbered pages 1040 and 1041.

and above all, not to 'bring in such a bill as at the present, which, without contributing possibly in the smallest degree to the object in view, will let loose the most dreadful scourge upon the lower orders of the people' (p.1027*).Windham's oratory was much the most powerful delivered in the Commons on this subject, and Wilberforce's attempt to appeal to his hearers' sensibility sounds from its report to have been thwarted by lack of time:

> Mr Wilberforce [said] ... he possessed a letter, which, if he read, he was sure would create in the breast of the right hon gent sentiments congenial to his own. ... By raising the estimation of the animal creation in the minds of the ignorant, this bill would create a sum of sensitive happiness almost impossible to calculate. (p. 1029*)

Sir Samuel Romilly cited in support of the bill Hogarth's well-known engravings of 1750–51 on *The Four Stages of Cruelty* to illustrate the traditional lesson that cruelty to animals begets cruelty to human beings, and concluded that 'He really believed this bill might be considered in a great degree as a bill for the prevention of cruel murders' (p. 1031*). Despite such lack-lustre performances from its supporters in the Commons, the bill was voted into Committee by 40 votes to 27, but on 19 June, at the very end of the session, the House reversed its decision by 37 votes to 27, and the bill was lost.

The social historian James Turner has perceived in Windham the articulator of a justified anxiety concerning the vast change which was sweeping, almost unnoticed, over early nineteenth-century Britain, and has shown how Windham's picture of a bucolic golden age, which he maintained the enemies of bull-baiting were trying to destroy, was – perhaps unconsciously – an attempt to stand fast against the loss to industrialization of 'Old England':

> Windham's lurid picture of conspiratorial goings-on was far-fetched, but he was right – perhaps more right than he and his opponents knew – to insist that the whole rural culture, the traditional agricultural way of life, was somehow at stake in the struggle over bull-baiting.[23]

Windham's project is in this respect not unlike Wordsworth's portrayal of the countryside as a 'hard pastoral' which, as I discuss further in Chapter Five, has little time for sentimentality about animals, seeing them instead as part of the basic fabric of agricultural life. Windham's allusions to Sterne, Goldsmith and Gray also indicate, however, the way in which – as the English countryside began to come under threat from industrialization in the late eighteenth and early nineteenth centuries – rural England became the site in political as well as literary discourse for themes of nostalgia and loss. In this context it is ironic that Windham should have singled out Lancashire and Staffordshire as the counties most active in bull-baiting and therefore as the source of the best soldiers for the army, since these were also two of the most industrialized areas of the country, with a working class that had already begun to leave behind the purely pastoral existence which Windham sought to safeguard. As Hobhouse's and Byron's

---

[23] James Turner, *Reckoning with the Beast: Animals, Pain, and Humanity in the Victorian Mind* (Baltimore and London: John Hopkins University Press, 1980) p. 28.

Childe Harold IV commentary on blood and circuses shows, there is a deep confusion here between cause and effect, and not only the debate about cruelty and courage in war but also that about pastoral innocence and industrialized consciousness can readily be turned inside out by sophistical arguments.

It is not possible to establish how much of the progress of Erskine's animal cruelty bill Byron followed after the debate in the Lords which he attended on 15 May. He was certainly in London during this period, although he left the capital for Falmouth and the start of his 'grand tour' on the day the bill was finally defeated. Byron's comments in both 1813 and 1821 on Windham's style of oratory are entirely consistent with the features of Windham's performance in opposing the bill in the House of Commons on 13 June 1809. Since Windham died while Byron was still abroad in 1810, there were few other opportunities for Byron to have heard him speak, and I think the balance of probability is that Byron did watch and hear Windham's performance from the visitors' gallery of the Commons, or at least read the newspaper reports on this part of the debate. My discussion of the bullfight and associated stanzas of *Childe Harold* I which follows therefore assumes that Byron knew Windham's style and arguments at first hand from the 1809 animal cruelty debate and that he, like Hobhouse (from the evidence of the note to *Childe Harold* IV) was familiar with the content of the 1800 and 1802 debates from having read accounts of them.

Whether or not Byron was present at the later debates in the Commons on the animal cruelty bill, it is evident that he was, in stanzas 65 to 86 of *Childe Harold* I, joining the parliamentarians whose debates I have outlined in drawing upon Burke's articulation of the distinction between conservatism and radical change and the alignment, in the *Reflections on the Revolution in France*, of these political polarities with certain respective packages of cultural symbols. In both the debates and the poem, the animal cruelty issues provide not only subject matter in themselves but also a pretext for arguments about human culture and politics on a much wider scale, aligned according to Burkean categories.

Using Burke

Some of the participants in this discourse (including Byron, Windham and Sheridan) were highly conscious of this aspect of their expression and used the Burkean materials in a sophisticated way for their own purposes. Others, such as Erskine, Wilberforce and their associates, appear to have been much less aware of the nature of the battleground into which they were sallying. Thus Erskine seems unwittingly to have increased Windham's hostility by insisting on the anti-Burkean *newness* of his anti-cruelty measure, while Wilberforce used terms connected with sentiment and sensibility apparently unaware that they had been captured by Burke and diverted to other ends:

> the age of chivalry is gone. That of sophisters, economists and calculators, has succeeded; and the glory of the Empire is extinguished for ever. ... The unbought grace of life, the cheap defence of nations, the nurse of manly sentiment and heroic enterprise, is gone! it is gone, that sensibility of principle, that chastity of honour, which felt a stain like a wound, which inspired courage whilst it mitigated ferocity, which ennobled whatever it

touched, and under which vice itself lost half its evil, by losing all its grossness. ... This mixed system of opinion and sentiment had its origin in the ancient chivalry; and the principle, though varied in its appearance by the varying state of human affairs, subsisted and influenced through a long succession of generations, even to the time we live in. If it should ever be totally extinguished, the loss I fear will be great. *(Reflections,* pp. 73–4)

I have shown how Windham's position as Burke's disciple is demonstrated throughout his speeches in both style and subject-matter. Byron's engagement with the *Reflections* in particular is most clearly stated in the 'Addition to the Preface' to *Childe Harold* cantos I and II, added to the fourth edition which was published in September 1812. In this Byron defends his protagonist Harold against the charge of being 'unknightly' by claiming that, on the contrary, the 'very indifferent character of the vagrant Childe' is entirely consistent with that of

the good old times, when 'L'amour du bon vieux tems, l'amour antique' flourished, [which] were the most profligate of all possible centuries. ... The vows of chivalry were no better kept than any other vows whatsoever, and the songs of the Troubadours were not more decent, and certainly were much less refined, than those of Ovid. – The 'Cours d'amour, parlemens d'amour ou de courtesie et de gentilesse' had much more of love than of courtesy or gentleness. ... Whatever other objection may be urged to that most unamiable personage Childe Harold, he was so far perfectly knightly in his attributes – 'No waiter, but a knight templar.' By the by, I fear that Sir Tristram and Sir Lancelot were no better than they should be, although very poetical personages and true knights 'sans peur', though not 'sans reproche'. ... So much for chivalry. Burke need not have regretted that its days were over, though Maria Antoinette was quite as chaste as most of those in whose honours lances were shivered, and knights unhorsed.[24]

Stanzas 65 to 86 of *Childe Harold* I illustrate the themes that are set forth in this part of the Preface: the bullfight is characterized by repeated images of chivalric tournaments and by an intensification of the Spenserian language which Byron uses – almost always with mocking intent – throughout the poem to make associations with mediaevalism and the age of chivalry. Byron's descriptions of mediaeval chivalry are also indebted to Walter Scott, especially *Marmion* (1808): Peter Manning (p. 174) points out that Scott had already shaped the 'connection between the form of metrical romance and conservative ideology', although Burke had of course made the connection earlier, in the *Reflections.* Byron knows that his readers will anticipate this link, and plays ironically upon their expectations:

---

[24] *BCPW*, II. 5–6. Peter J. Manning points out (in 'Childe Harold in the Marketplace: From Romaunt to Handbook', in *Modern Language Quarterly* LII (1991) pp. 170–90 (p. 173) that the *British Review* had pronounced that Childe Harold was 'no child of chivalry'.

The lists are op'd, the spacious area clear'd,
Thousands on thousands pil'd are seated round;
Long ere the first loud trumpet's note is heard,
Ne vacant space for lated wight is found: ...

Hush'd is the din of tongues – on gallant steeds
With milk-white crest, gold spur, and light-pois'd lance,
Four cavaliers prepare for venturous deeds,
And lowly bending to the lists advance;
Rich are their scarfs, their chargers featly prance:
If in the dangerous game they shine to-day,
The crowd's loud shout and ladies' lovely glance,
Best prize of better acts, they bear away,
And all that kings or chiefs e'er gain their toils repay.
(I. 720–37)

The fact that this tournament-like scene is set in Spain also calls forth Burkean references. As I have shown, sympathetic treatment of mediaevalism, aspects of Roman Catholicism, and Catholic customs and traditions was part of Burke's argument, and Windham carried this forward in the 1800 bull-baiting debate when he contrasted British puritanism disparagingly with the customs of the south of France and Spain. The southern setting, and this subject matter, also links Burke's, Windham's and Byron's scenes with the Iberian romances, defined by Johnson in his *Dictionary* as 'military fable[s] of the middle ages; tale[s] of wild adventures in war and love.' Versions of these were widely read in both 'high' and 'low' forms in eighteenth century England. Alongside literary works such as Ariosto's *Orlando Furioso,* Spenser's *Faerie Queene,* 'the old Spanish romance of *Felixmarte of Hircania'* chosen by Johnson from Thomas Percy's library, and translations and revisions of *Don Quixote* which reworked the genre, they continued to appear in heavily curtailed chap-book versions of their original form, as the 'wild tales' which were preferred by Wordsworth, Coleridge and their circle to the new instructional children's books of the end of the century.[25]

The central theme of romance is the knight's quest for and acquisition of a lady by rescuing her from danger, involving – as in the case of St George – the killing of a dragon or wild beast in a fight told with much gory detail. Burke's chivalric invocation of Marie Antoinette's need for protection, and her power to act as an inspirational force, reflects this aspect in more decorous terms. There is, however, as Eithne Henson describes it (p. 20), 'an ethos of slaughter and sexuality' in these tales. Henson points out (p. 51) not only how women primarily appear in these male-authored narratives as commodities, but also how moral responsibility for male aggression is displaced onto them: 'Since it is the lady who sends the knight out to gain an honourable name before she will accept

---

[25] Thomas Percy, quoted in *Boswell's Life of Johnson* ed by R. W. Chapman (Oxford: Oxford University Press, repr., 1985) p. 36. *Don Quixote* was popular enough to have been read by five out of Hester Thrale's eighteen servants: 'No book was ever so popular as Don Quixote; the Classics themselves are more confined in Fame: Don Quixote is the Book for high & low.' *Dr Johnson by Mrs Thrale: the 'Anecdotes' of Mrs Piozzi in their Original Form,* ed by Richard Ingrams (London: Chatto & Windus, 1984) p. 108.

his love, she can be considered morally responsible for the violence that follows.'

Byron's subtitle to *Childe Harold's Pilgrimage* – 'a Romaunt' – indicates how his treatment of the bullfight and of Cadiz in general uses Burkean ideas and symbols as what Peter Manning calls 'the locus for continuing debate over the claims to legitimacy of the *ancien régime;* ... a contested site, given immediacy by the language invoked to hold Britain's resistance to Napoleon' (p. 170). Byron deploys this cultural package of romance, chivalry and nostalgic conservatism satirically both to point up the horrors of the real war that was ravaging the Iberian peninsular in 1809, and also to explore the behaviour and role of women in a foreign, specifically southern, society and during war-time. The bullfight focuses attention on these issues through a scene which is a microcosm of war and of a society at war, using the enclosed arena as synecdoche for the wider field of combat. The surrounding stanzas introduce and establish the themes which are treated there in an especially concentrated form.

Chivalry is not long in appearing by name in *Childe Harold* I (stanza 24) and is first invoked as a 'goddess' in stanza 37, indicating the way in which femaleness in the following stanzas will be allied with slaughter and how chivalric references will almost always be ironically contrasted (and thus in fact intimately connected) with the bloody business of contemporary warfare:

> Awake, ye sons of Spain! awake! advance!
> Lo! Chivalry, your ancient goddess, cries,
> But wields not, as of old, her thirsty lance,
> Nor shakes her crimson plumage in the skies:
> Now on the smoke of blazing bolts she flies,
> And speaks in thunder through yon engine's roar.
> (I. 405–10)

The chivalric references are carried forward in the description of the battle of Talavera, fought on 27 and 28 July 1809:

> By Heaven! it is a splendid sight to see
> (For one who hath no friend, no brother there)
> Their rival scarfs of mix'd embroidery,
> Their various arms that glitter in the air! ...
> All join the chase, but few the triumph share;
> The Grave shall bear the chiefest prize away,
> And Havoc scarce for joy can number their array.
> (I. 432–40)

The failure of chivalry as a code of conduct for the aristocracy in conducting modern war is made explicit in the stanza in which the poet bids farewell to Cadiz (at this time the only city in Spain where bullfights were still permitted):

> Here all were noble, save Nobility;
> None hugg'd a conqueror's chain, save fallen Chivalry! (I. 880–1)[26]

---

[26] Sir John Carr and Byron attended the same bullfight, on 30 July 1809, at Puerto Maria, near Cadiz. Sir John opens his account of the fight, in his *Descriptive Travels in the Southern*

The bullfight that Byron saw took place on a Sunday, and this too becomes an opportunity for anti-Burkean satire. The introduction to the scene ironizes the Protestant's shock at such a use of the Sabbath:

> What hallows it upon this Christian shore?
> Lo! it is sacred to a solemn feast. ...
>
> Soon as the matin bell proclaimeth nine,
> Thy saint adorers count the rosary:
> Much is the VIRGIN teaz'd to shrive them free
> (Well do I ween the only virgin there)
> From crimes as numerous as her beadsmen be;
> Then to the crowded circus forth they fare,
> Young, old, high, low, at once the same diversion share.
>   (I. 685–719)

The strangeness of this Spanish 'day of blessed rest' is contrasted with a London 'day of prayer' – not, however, with one that is any more pious than its southern counterpart. Instead, in what Byron called 'two Stanzas of a buffooning cast' (69–70), he describes how London's 'spruce citizen, wash'd artizan, / And smug apprentice' leave the capital by various means to 'gulp their weekly air', and in particular how 'many to'the steep of Highgate hie':

> Ask ye, Boeotian shades! the reason why?
> 'Tis to the worship of the solemn Horn,
> Grasp'd in the holy hand of Mystery,
> In whose dread name both men and maids are sworn,
> And consecrate the oath with draught, and dance till morn.
>   (I. 706–10)

Byron offered in 1811 to 'delete these two stanzas before publication, and Thomas Moore thought in 1832 that they continued to 'disfigure the poem' *(WLBP,* II. 65). However, they are part of the wider scheme of the work in that they also offer a critical commentary upon Burke's invocation of tradition as the basis of good government and civilized society: the way in which, according to him, 'We procure reverence to our civil institutions on the principle upon which nature teaches us to revere individual men; on account of their age, and on account of those from whom they are descended' *(Reflections,* p. 32). The point about Byron's London Sunday activities is that the 'traditions' they invoke are spurious ones: especially the custom of 'swearing on the horns' at Highgate. Hone's *Everyday Book* of 1827 says that 'Drovers, who wished to keep the tavern to themselves, are said to have been responsible for the rude beginnings of this tedious foolery,' and quotes this sample of the oath that customers had to take in the presence of the Landlord before they could drink there:

---

and Eastern Parts of Spain and the Balearic Isles, in the Year 1809 (London: Sherwood, Neely and Jones, 1811), with the statement (p. 53) that 'the only bull-fight in Spain was at this time here', and he describes how this, the Spaniards' 'favourite pastime', had been officially 'abolished' by Charles IV in 1805 (p. 54).

'You must not eat brown bread where you can get white, except you like the brown best. You must not drink small beer where you can get strong, except you like the small best. You must not kiss the maid while you can kiss the mistress, but sooner than lose a good chance you may kiss them both.' *(WLBP,* II. 66)

The attempt of traditionalists such as Windham to identify customs like bull-baiting as quintessentially rural and, on the grounds of their age and supposed rural nature, with 'the genius and spirit of the country', is undermined by passages like this which associate similar customs with urban rather than rural life, and with possibly mildly lewd behaviour, but mainly with a bogus, meaningless and 'tedious foolery'.

Women at war

Susan Wolfson and Caroline Franklin have explored at length Byron's complex portrayal of women in *Don Juan* to show how Byron adheres to an image of passionate, irrational and 'natural' femininity in antithesis to Wollstonecraft's model of the rational and passionless woman who rivals or surpasses in essentially masculine qualities the male citizen of liberal, republican discourse.[27] Franklin shows how, in *Don Juan,* as part of Byron's 'twin themes: the satire of insatiable female passion and the denunciation of despotism' ('Juan's Sea-Changes', p. 73):

> [t]he poem's experimentation with sexual role-reversal implies that conventional 'femininity' and 'masculinity' are culturally constructed and subject to change. Though Wollstonecraft finds nothing to admire in conventional femininity, Byron makes his hero transgress the traditional boundaries of gender in order to explore feminine subjectivity and experience, as well as inscribing them in female characters, because of their subversive potential. Though femininity is still conventionally portrayed as irrational and volatile, it is also romanticized as closer to nature than 'rational' masculinity. ('Juan's Sea Changes', p. 84)

These stanzas of *Childe Harold* I, which reflect a youthful and relatively inexperienced Byron's first meeting with foreign women and his first close encounter with war, show many of the same features in place as early as 1809. Byron's evocation in stanza 81 of the carefree pleasures of the south resonates with Windham's portrayal in the 1800 bull-baiting debate of how in Spain 'in the cool of the evening's shade, the poor dance in mirthful festivity on the green, to the sound of the guitar':

---

[27] Susan J. Wolfson, '"Their she condition": Cross-dressing and the Politics of Gender in *Don Juan*', in *Journal of English Literary History,* LIV (1986) pp. 585–617. Caroline Franklin, *Byron's Heroines* (Oxford: Clarendon Press, 1992); and 'Juan's Sea Changes: Class, Race and Gender in Byron's *Don Juan'* in *Don Juan* ed by Nigel Wood (Buckingham: Open University Press, 1993) pp. 56–89.

Who late so free as Spanish girls were seen,
(Ere war uprose in his volcanic rage),
With braided tresses bounding o'er the green,
While on the gay dance shone Night's lover-loving Queen?
(I. 806–9)

However, Byron's ambiguous use of the term 'free' to describe the Spanish
girls, and the presence of the moon as a presiding 'lover-loving' goddess, points
back to the parody with which Byron treats the ladies of chivalry in the
'Addition to the Preface' to *Childe Harold* I and II. His remarks there about
Marie Antoinette specifically eschew Burke's chivalrous evocation of the Queen
as 'glittering like the morning-star, full of life, and splendour, and joy', and
imply that she may have been adulterous. The association of southern, hotter
climates and Roman Catholicism with sensuality and passion – especially female
passion – is one that Burke conspicuously does *not* make, but it is a topic
which, following Montesquieu's *The Spirit of the Laws* (1748) was commonly
referred to by British writers in connection with travels in warmer climates.[28]
Northern European male writers travelled to the south ready to find female
sensuality, and to Spain in particular primed with expectations of a continuing,
though debased, chivalric romanticism, and they generally found what they
sought.

Sir John Carr, for example, who was in Cadiz at the same time as Byron,
observed:

> The insensibility of that man must be great indeed, who cannot find
> a querida, or one to whom he is permitted to devote all his soul,
> amongst either the married or the unmarried; and destitute of
> every attraction must that woman be, who does not meet with a
> cortija or lover, or rather her impassioned slave, amongst the men.
> In carrying on an intrigue, the Spanish ladies are singularly
> dexterous. Wrapped up in the masquerade of fable and parable,
> they carry on an amorous conversation with their admirers in
> public, without fear of detection. *(Descriptive Travels,* p. 10)[29]

The mature Byron went on to treat the subject with the assurance of irony:

---

[28] See Charles de Secondat, baron de Montesquieu, *The Spirit of the Laws* trans. and ed
by Anne M. Cohler, Basia Carolyn Miller and Harold Samuel Stone (Cambridge: Cambridge
University Press, 1989) p. 233: 'In northern climates, the physical aspect of love has scarcely
enough strength to make itself felt; in temperate climates, love, accompanied by a thousand
accessories, is made pleasant by things that at first seem to be love but are still not love; in
hotter climates, one likes love for itself; it is the sole cause of happiness, it is life.' See also
Part 3, pp. 231–84: 'On the Laws in their Relation to the Nature of the Climate'; 'How the
Laws of Civil Slavery are Related to the Nature of the Climate; 'How the Laws of Domestic
Slavery are Related to the Nature of the Climate,' and 'How the Laws of Political Servitude are
Related to the Nature of the Climate.'

[29] An unpublished variant of stanza 87 refers to Carr:
Ye! who would more of Spain and Spaniards know ...
Go hie ye hence to Paternoster Row,
Are they not written in the Boke of Carr,
Green Erin's knight! and Europe's wandering star! ...
This borrow, steal (don't buy) and tell us what you think.

What men call gallantry, and gods adultery,
Is much more common where the climate's sultry;
(Don Juan I: 63 [1818])

and by the time he was writing to Hobhouse from Ravenna in July 1821, somewhat defensively to explain why at the request of his Italian mistress Countess Teresa Guiccioli he had left off writing *Don Juan,* he had elaborated a theory about the relationship between chivalry and sexuality well suited to his own purposes. Included in that pàckage is a contrast between a Burkean admiration for the Gothic and Byron's clear preference for classicism and Hellenism.[30] (The same preference was already being stated by Byron in 1812, in his claim in the 'Addition to the Preface' of *Childe Harold* I and II that 'the songs of the Troubadours weré not more decent, and certainly were much less refined, than those of Ovid'.) Of the Gothic Burke had written:

> It is this which has given its character to modern Europe. It is this which has distinguished it under all its forms of government, and distinguished it to its advantage, from the states of Asia, and possibly from those states which flourished in the most brilliant periods of the antique world *(Reflections,* p. 74)

while of Teresa's dislike of *Don Juan* Byron commented:

> She had read the French translation and thinks it a detestable production. – This will not seem strange even in the Italian morality – because women all over the world always retain their Free masonry – and as that consists in the illusion of the Sentiment – which constitutes their sole empire – (all owing to Chivalry – & the Goths – the Greeks knew better) all works which refer to the *comedy* of the passions – & laugh at Sentimentalism, of course are proscribed by the whole *Sect.* ... You will be very glad of this – as an earlier opponent of that poem's publication. *(BLJ* VIII. 148)

In *Childe Harold* I, however, the subject of southern female sexuality is still approached with some shock and fascination. The city of Cadiz (then pronounced by English-speakers in a way which made its obvious rhyme 'ladies'[31]) and its inhabitants are presented as ruled over by the goddess of love, who has supplanted even the supposedly lax ways of Roman Catholicism in her worship:

---

[30] 'Gothic theory' had a long history in British (and French) political thought, going back to the seventeenth and sixteenth centuries. For Scots and for Tories, it could stand for northern patriotism, Jacobitism, and an opposition to the perceived corruption of Walpole's Whig government. In England, however, it had been primarily a Whig concept invoking supposed 'Saxon' and pre-Norman models of free government as a means of validating political systems that curtailed the power of the monarch. In his use of 'the Gothic' to oppose the French Revolution, Burke can be seen to have 'transmuted the Whig historical tradition and bequeathed it, much changed, to his nineteenth-century conservative successors': see R. J. Smith, *The Gothic Bequest: Mediaeval Institutions in British Thought, 1688–1863* (Cambridge: Cambridge University Press, 1987) p. 113.

[31] See *Don Juan* II, stanza 81.

> But Cadiz, rising on the distant coast,
> Calls forth a sweeter, though ignoble praise.
> Ah, Vice! how soft are thy voluptuous ways!
> While boyish blood is mantling, who can 'scape
> The fascination of thy magic gaze? ...
>
> When Paphos fell by Time – accursed Time!
> The queen who conquers all must yield to thee –
> The Pleasures fled, but sought as warm a clime;
> And Venus, constant to her native sea,
> To nought else constant, hither deign'd to flee;
> And fix'd her shrine within these walls of white:
> Though not to one dome circumscribeth she
> Her worship, but, devoted to her rite,
> A thousand altars rise, for ever blazing bright. ...
>
>   A long adieu
> He bids to sober joy that here sojourns:
> Nought interrupts the riot, though in lieu
> Of true devotion monkish incense burns,
> And Love and Prayer unite, or rule the hour by turns.
>   (I. 659–83)[32]

It is within this setting – of the dominion of Venus – that the description of the bullfight is foregrounded. The solemn feast to which the Sabbath is 'hallowed' is the bullfight, and this, like the other rites which have displaced 'real' religion, is also presided over by women:

> Here dons, grandees, but chiefly dames abound,
> Skill'd in the ogle of a roguish eye,
> Yet ever well inclin'd to heal the wound;
> None through their cold disdain are doom'd to die
> As moon-struck bards complain, by Love's sad archery.
>   (I. 724–8)
>
> The throng'd Arena shakes with shouts for more;
> Yells the mad crowd o'er entrails freshly torn,
> Nor shrinks the female eye, nor ev'n affects to mourn.
>   (I. 689–92)

As in the depiction of the warlike, though beautiful and gentle-seeming, Maid of Saragoza in stanzas 54–8, who 'all unsex'd the Anlace hath espous'd, / Sung the loud song, and dar'd the deed of war', the poet meditates here on the conjunctions he perceives in southern women between aggression and sexuality, and between passionate love and cruelty, by referring these to the myths of chivalry. Windham had sought in the 1809 debate to show how even the most sophisticated women of the northern world, who prided themselves on their sensibility and their support for animal protection, were implicated in fashionable practices which harmed animals. Like Windham and the writers of

---

[32] Byron's attitude to Roman Catholicism later mellowed to the point of apparent approval: see *BLJ* IX: 123 (1822).

the romances, Byron transfers responsibility to women for – and then charges them with particular hypocrisy over – the cruelty of which he disapproves, although men are its actual protagonists.[33]

Two of Byron's English-speaking companions at the bullfight expressed a similar mixture of repellence and fascination about women's enthusiastic participation in the sport as spectators.[34] Sir John Carr titillates his readers' interest by describing the level of excitement raised when a young marquis, who had 'already won several ladies' hearts by his beauty, and his prowess' joined the matadors in the ring: '"Oh what merit has that fine young nobleman," said a pretty Spanish lady, "how beautifully did he kill the bull!"' (Descriptive Travels, p. 61). Carr is perhaps deliberately unclear about whether the glamour and sexual interest in the sport is generated by the human or the animal participants – by the skill or by the slaughter – when he records how the bulls are also lavishly rewarded with female attention: 'the governor's daughter had honoured the beast by making with her own delicate hands, a rich decoration of ribbons for his neck, and lovely women applauded the bloody havoc which he made' (p. 61); or when he knowingly remarks how 'every Spanish lady is as well acquainted with all the fine points of a bull, as an English one is of those of a lap-dog' (p. 65).[35]

Byron's travelling companion Hobhouse had the benefit of a later viewpoint and many further years' experience of foreign travel when he described more coolly in a note to Childe Harold IV in 1818, how:

> The wounds and death of the horses are accompanied with the loudest acclamations, and many gestures of delight, especially from the female portion of the audience, including those of the gentlest blood. Everything depends on habit. ... A gentleman present, observing [us] shudder and look pale, noticed that unusual reception of so delightful a sport to some young ladies, who stared and smiled, and continued their applause as another horse fell bleeding to the ground. ... An Englishman who can be much pleased with seeing two men beat themselves to pieces, cannot bear to look at a horse galloping round an arena with his bowels trailing on the ground, and turns from the spectacle and the spectators with horror and disgust. (BCPW II 258–9)[36]

---

[33] Women now participate in bullfighting as fully-fledged matadors. Tim Brown reported in the Daily Telegraph on 13 May 1998 that the first 'doctorate' of the profession to be conferred on a woman was received the day before by 26-year-old Cristina Sanchez in the Las Ventas bullring in Madrid. Ms Sanchez killed two bulls on that occasion: she had previously fought bulls more than 200 times and had been seriously gored three times. Women's participation on foot in the bullring became legal in 1974.

[34] The conflict between women's role as objects for the male gaze and as spectators themselves has been a hotly-debated area in contemporary film and other theory: see, for example Laura Mulvey's 'Visual Pleasure and Narrative Cinema' in Screen XVI, 3 (Autumn 1975) pp. 6–18, where Mulvey shows how 'the determining male gaze projects its phantasy on to the female figure which is styled accordingly ... coded for strong visual and erotic impact so that they can be said to connote to-be-looked-at-ness' (p. 11).

[35] Thomson's criticism of women's participation in hunting in The Seasons was well-known: 'Let not such torrid joy / E'er stain the bosom of the British fair. / Far be the spirit of the chase from them!' ('Autumn', 571–3).

[36] There is probably a dig at Byron here, since he had been an aficionado of pugilism.

As Hobhouse points out, the sufferings of horses – which were for English gentlemen of this period a close adjunct to their society and thus came into the category of what Lévi-Strauss was to call 'metonymical human beings' (see further below, Chapter Four) – might actually be more painful to this group of spectators than those of men.[37] Byron's description of the scene draws special attention to the bond between horse and rider: valorizing the animal, as he does the old horse in 'Mazeppa' and the dog in his 'Inscription' for the Newfoundland, in terms of the service it renders to its human 'lord', and thus magnifying the reader's perception of the guilt of the animal's tormentors: 'Alas! too oft condemn'd for him [man] to bear and bleed (I. 746):

> One gallant steed is stretch'd a mangled corse;
> Another, hideous sight! unseam'd appears,
> His gory chest unveils life's panting source,
> Tho' death-struck still his feeble frame he rears,
> Staggering, but stemming all, his lord unharm'd he bears.
>   (I. 769–73)

Finding a hero

As a narrative of human inhumanity, recounted in the present tense and with minute particulars, Byron's description of the death of the horses – and particularly that of the bull – are similar in type and scale to those which advocates of the anti-cruelty measures used in Parliament:

> Foil'd, bleeding, breathless, furious to the last,
> Full in the centre stands the bull at bay,
> Mid wounds, and clinging darts, and lances brast,
> And foes disabled in the brutal fray;
> And now the Matadores round him play,
> Shake the red cloak, and poise the ready brand:
> Once more through all he bursts his thundering way –
> Vain rage! the mantle quits the conynge hand,
> Wraps his fierce eye – 'tis past – he sinks upon the sand!
>
> Where his vast neck just mingles with the spine,
> Sheath'd in his form the deadly weapon lies.
> He stops – he starts – disdaining to decline:
> Slowly he falls, amidst triumphant cries,
> Without a groan, without a struggle dies.
>   (I. 774–87)

What the parliamentarians did *not* do, however, and Byron most deliberately does, is to make the bull the hero of the scene. The reader's sympathy for him is called forth not by pity or sensibility, as in the parliamentarians' accounts, nor by guilt, as in Byron's own description of the horses' death, but by admiration for the bull's courage and dignity. In this setting it is perhaps *only* animals

---

[37] See Claude Lévi-Strauss, *The Savage Mind* (London: Weidenfeld and Nicolson, 1966) p. 205.

which can stand outside the blood-guilt which disfigures every human hand. 'I want a hero,' Byron claims at the opening of *Don Juan*, going on to detail why the Napoleonic wars, for all their bloodshed, have not provided the kind of hero he needs: and, despite the gloomy eponymous Childe, the lack is there in the *Pilgrimage* also. In these stanzas Byron reverses the scenes of chivalric romance where the human hero must fight a dragon, wild beast or monster to prove himself and win his lady, so that it is the bull's valour, rather than that of his human opponents, which is foregrounded.

It is not insignificant that this scene foreshadows the 'dying gladiator' stanzas in *Childe Harold* IV (139–42), where the reader is made to feel as if singled out from the surrounding crowd in order to be given special awareness of the individual sufferings of a lone, isolated figure.

> I see before me the Gladiator lie:
> He leans upon his hand – his manly brow
> Consents to death, but conquers agony,
> And his drooped head sinks gradually low –
> And through his side the last drops, ebbing slow
> From the red gash, fall heavy, one by one,
> Like the first of a thunder-shower; and now
> The arena swims around him – he is gone,
> Ere ceased the inhuman shout which hail'd the wretch who won.
> *(Childe Harold* IV. 1252–60)

The gladiator is, of course, a creature of Byron's imagination, based upon a sculpture in the Capitoline museum, and it appears that the bull's death-scene is a similar fiction since, judging by Carr's and Hobhouse's accounts, it seems unlikely that Byron actually saw a bull die in the fight they attended on 30 July 1809. The spectacle instead consisted in a very fierce and experienced bull, apparently belonging to a priest, killing several horses with its horns.[38] As with the gladiator's, so the bull's death is played out as a scene in Byron's imagination. From the generality of the description of the crowd – 'Gapes round the silent Circle's peopled walls' (I. 749) – the poet focuses in on small details which encourage the reader to do what Byron has already done, by identifying with the bull imaginatively and in terms of feeling:

> He flies, he wheels, distracted with his throes;
> Dart follows dart; lance, lance; loud bellowings speak his woes.
> (I. 763–4)

When the dagger enters the bull's body, we are told precisely at which spot – 'Where his vast neck just mingles with the spine' – and the pinching out of the vowels towards the end of the line seems to concentrate the sensation so that, like the bull, the reader might sense the sharp point entering the nape of the neck.

The invocation of such sensitivity or sensibility was, as has been shown, one of the primary tactics of the parliamentary supporters of the anti-cruelty measures. Byron's project, like theirs, uses accepted motifs of and arguments

---

[38] Hobhouse's *Diary* for Sunday 30 July 1809 records '4 horses kill'd by one black bull (a priest's)'.

about cruelty, such as this, to present its case, and I believe the evidence indicates that Byron viewed the Spanish bullfight from the beginning in terms constructed by the British parliamentary debates, and then combined this theoretical introduction with his own experience of observing cruelty first hand, to meet his own specific ends. The first point made and very often repeated by the Wilberforcean politicians – that cruelty to animals degrades human beings – is a mainspring of Byron's argument. The bullfight is a microcosm for a country at war in two distinct ways. First, it illustrates in a new way the old, Thomist and Hogarthian argument about how human beings are made cruel towards each other by practising and watching cruelty to animals, as the bullfighting cult deliberately trains up Spaniards of both sexes for bloodshed:

> Such the ungentle sport that oft invites
> The Spanish maid, and cheers the Spanish swain.
> Nurtur'd in blood betimes, his heart delights
> In vengeance, gloating on another's pain.
> (I. 792–5)

The perception of parallels between Byron's dying gladiator and his dying bull is clarified by Hobhouse's acerbic comment on the Coliseum passage *(Childe Harold* IV. 1267) about 'Mr Windham's panegyric on bull-baiting', and lines 792–5 of *Childe Harold* I may be read as a direct riposte to Windham's elaboration of the idea that sports such as bull-baiting were not only desirable but necessary in the national interest, 'to produce the best soldiers for the army'. Both sets of verses illustrate Byron's belief that cruelty – to animals or to slaves – indicates degeneracy in the people who practise it, whether Roman emperors or a debased Spanish nobility. Stanza 80 indeed goes on to show why such sports actually vitiate a people's ability to fight effectively for their country, by demonstrating how the bloodthirstiness induced by participation in blood-sports is turned into local feuding instead of united and effective opposition to the invasion of the country:

> What private feuds the troubled village stain?
> Though now one phalanx'd host should meet the foe,
>   Enough, alas! in humble homes remain,
> To meditate 'gainst friends the secret blow,
> For some slight cause of wrath, whence life's warm stream must flow.
> (I. 796–800)

The second means by which the bullfight scene works to characterize a country at war lies in its exploration of the way human beings actively disguise from themselves the reality of slaughter by the chivalric glamour they cast over bloodshed. In the bullring the animals' lack of guilt throws this into relief. The bull is innocent – despite the deaths he causes and the injuries he inflicts – partly because he is provoked by human beings, but particularly because his ferocity is natural: a part of the inherent wild energy and aggression (he is 'the forest-monarch' and 'the lord of lowing herds') that Byron so often notices in animals but accepts uncritically, and even relishes as an intrinsic part of the essential truth and truthfulness of their nature.[39] Like Sheridan, and following Rousseau, he is an admirer of 'animals in their natural state and exhibiting their natural qualities'. Human beings, on the other hand, possess reason, and therefore have the capacity for deluding and lying to themselves about their savagery: in the words of the 'Inscription' for the Newfoundland Dog, while the dog has an 'honest heart' (9), Man – who is *animated* dust' (18, my italics) and therefore has the capacity for reason – leads himself into being 'Debased by slavery, or corrupt with power' (16), and therefore becomes 'disgusting' and 'degraded' (17–18). For Byron as for Erskine and for Bentham, the capacity for pain and pleasure which links animals with human beings makes the possibility of rights for animals an option to be taken seriously. For Byron as for Wilberforce, it is the human beings who are 'brutalized' by violence and bloodshed, specifically because of their possession of reason: but whereas for Wilberforce human cruelty to animals makes human beings risk being reduced to animals themselves ('degrad[ing] human nature to a level with the brutes'), for Byron the unreasoning animals – despite their ferocity – remain not only innocent but also morally superior to their human tormentors.

Political cover-up

One of the main themes of *Childe Harold* I – from the foolscap folly of the British at the Convention of Cintra (stanzas 24–5), to the Spanish 'royal wittol Charles' (stanza 48), and the 'bloated Chief's unwholesome reign' (stanza 53) of Napoleon – is the self-delusion practised by all sides in the Peninsular War. Byron sees human beings, especially war-leaders, as cloaking, disguising or throwing a veil of sentiment and fine language over their brutality, and he reads Burke's invocation of chivalry, with its repeated images of clothing, draping and covering up, as the direct advocacy of such obfuscation:

> All the pleasing illusions, which made power gentle and obedience liberal, which harmonized the different shades of life, and which, by a bland assimilation, incorporated into politics the sentiments which beautify and soften private society, are to be dissolved by this new conquering empire of light and reason. All the decent drapery of life is to be rudely torn off. All the superadded ideas, furnished from the wardrobe of moral imagination, which the heart

---

[39] Erskine associated liberty with wild animals in the famous peroration to his speech in defence of Stockdale in the Constructive Treason trials of 1794: 'It is the nature of every thing that is great and useful, both in the animate and inanimate world, to be wild and irregular – and we must be contented to take them with the alloys which belong to them, or live without them. ... Liberty herself, the last and best gift of God to his creatures, must be taken just as she is; – you might pare her down into bashful regularity, and shape her into a perfect model of severe scrupulous law, but she would then be Liberty no longer; and you must be content to die under the lash of this inexorable justice which you had exchanged for the banners of freedom.' Quoted in Malcolm Kelsall, *Byron's Politics* (Sussex: Harvester Press, 1987) p. 17.

owns, and the understanding ratifies, as necessary to cover the
defects of our naked, shivering nature, and to raise it to dignity in
our own estimation, are to be exploded as ridiculous, absurd, and
antiquated fashion. ... On this scheme of things, a king is but a
man, a queen is but a woman; a woman is but an animal not of the
highest order. All homage paid to the sex in general as such, and
without distinct views, is to be regarded as romance and folly.
(Reflections, p. 74)

The subject of the 1800, 1802 and 1809 animal cruelty debates was only ever
extremely subsidiary to what was perceived as the proper business of both
Houses of Parliament during this period – 'beneath the deliberate dignity of
parliament; especially in times like the present, when questions of vital
importance are hourly pressing on our attention,' as Windham said of the bill of
1800 *(History,* XXXV. 204). For that very reason, however, the content of
these debates offered more possibilities for individualized ways of participation
in them, and greater opportunities for imaginative extrapolation from their
subject-matter, than did the regular parliamentary sessions of the time which
covered subjects such as, for example, the union with Ireland, the war with
France, attempts to prevent adultery and limit the duration of income tax, a bill
against treason and one on insane offenders *(History,* XXXV. 393 and *passim).*
     Byron was right when he told Medwin that he was 'not made for what you
call a politician': his own parliamentary speeches were too conspicuously
polished, mannered, metaphorical and literary in character to operate effectively
as political interchange.[40] When he wrote *Childe Harold* I and II, however, he
was still seriously interested in making his mark in Parliament: regarding his
literary 'scribbling' – as he was to continue to do at times throughout his life –
as secondary to the possibilities of a life of political action; and the swift fame
which *Childe Harold* I and II acquired, especially in Whig circles, was partly
due to the inclusion of overt political commentary alongside the imaginative
material in the verse. The animal cruelty debates, including the one he heard just
before he went abroad, provided Byron with ideas about animals which were
already packaged in terms of different political discourses (including those
connected with the Society for the Suppression of Vice's concern about the
brutalizing effects of cruelty; with Utilitarianism, and with Burke's material on
conservatism and chivalry) which he could combine with imaginative material to
achieve his own particular purposes. The debates also drew upon literary motifs
and the subject-matter of literary genres such as pastoral in order to effect their
aims; and this chapter has sought to show how the political and the literary
discourses both interrogate and help to develop a common culture about animals,
and how closely the two were connected and interchanged. The way in which
another kind of non-literary discourse – that connected with early
'vegetarianism' (the practice was not named as such until the 1840s) – is
deployed in Romantic-period poetry in connection with animals will be
discussed in the next chapter.

---

[40] *Medwin's Conversations,* p. 228. See also Andrew Nicholson's assessment of Byron's
speeches in *BCMP* p. 279.

Chapter 4

Animals as Food:
Shelley, Byron and the Ideology
of Eating

> then we lingered not,
> Although our argument was quite forgot;
> But, calling the attendants, went to dine
> At Maddalo's; yet neither cheer nor wine
> Could give us spirits, for we talked of him,
> And nothing else, till daylight made stars dim.
> (*Julian and Maddalo,* 519–24)[1]

It is a pity we never learn what Julian and Maddalo actually ate for dinner that memorable evening, and still more so that their preoccupation with the Maniac prevents us from hearing them discuss their dinner, or dinners in general. Given the intense relationship with food, eating and diet sustained by both Shelley and Byron, the subject might have made a good topic for the semi-autobiographical conversation and argument of the poem. Shelley might have deployed his self-image in Julian as a politically-motivated, idealistic partaker of food, and Maddalo/Byron would no doubt have responded with impatience at such 'Utopian' talk on a subject so resistant to meliorism.[2]

The readers of this imaginary piece of verse would not, I think, have been invited by Shelley to *share* in the 'cheer' and 'wine' which his protagonists consume, by means of descriptions which 'awaken ... in others, sensations like those which animate [the poet]'.[3] Keats's willingness to 'burst joy's grape against his palate fine'; his 'lucent syrops tinct with cinnamon'; his 'blushful

---

[1]   *Shelley, Poetical Works,* ed by Thomas Hutchinson (Oxford: Oxford University Press, 1970).

[2]   'We might be otherwise – we might be all
    We dream of happy, high, majestical.
    Where is the love, beauty and truth we seek
    But in our mind? and if we were not weak
    Should we be less in deed than in desire?'
    'Aye, if we were not weak – and we aspire
    How vainly to be strong!' said Maddalo:
    'You talk Utopia.'
    (*Julian and Maddalo,* 172–9)
Julian seems to drink wine, unlike Shelley, who through most of his life abstained from alcohol, and who wrote in a note to *Queen Mab* in 1813: 'the use of animal flesh and fermented liquors directly militates with this equality of the rights of man'. However, in the Preface to *Julian and Maddalo*, Shelley recorded of Maddalo/Byron: 'His more serious conversation is a sort of intoxication; men are held by it as by a spell,' and the reference to wine in the poem may be metaphorical rather than literal.

[3]   Preface to *Queen Mab.*

Hippocrene', and even his 'savour of poisonous brass and metal sick', are
outside the range in which Shelley wished to operate.[4] Both Shelley and Byron
would have concurred with at least the first two-thirds of Hegel's contention
that 'smell, taste and touch remain excluded from the enjoyment of art', in so
far as this can be taken to refer to attempts, like Keats's, to render the sensuous
impressions of taste and smell through the medium of words.[5] Both the older
poets give far more attention to the meaning, symbolism and ideology of this
form of consumption than to the evocation of the sensuous pleasures of the act
itself, and common to Shelley and Byron is an awareness of the connection
between contemporary discourses on food and eating and the radical politics of
tyranny, power and freedom, and a wish to use and explore this in their
writing.

A key part of these ideologies concerns the place of meat in human diet, and
the relationship this predicates between human beings and animals. This
chapter, therefore, carries forward an exploration of the text and context of
animals in Romantic-period writing by considering Shelley's idealistic non-
violent 'vegetarianism' and Byron's troubled carnivorousness in relation to
contemporary discourses about meat-eating and diet, and discusses how the
poets deploy these in their imaginative work.[6] It draws upon the separate
studies which there have been in this area: Timothy Morton, in *Shelley and the
Revolution in Taste*, has very thoroughly examined Shelley's conception of
how the body, perceived through its consumption of food, is 'an interface
between society and the natural environment'; Peter Graham has discussed the
order and disorder of eating as a means of self-empowerment in Byron's *Don
Juan*, and Wilma Paterson and others have suggested that Byron's writing
about food and his personal eating habits indicate that he suffered from
anorexia nervosa and bulimia nervosa.[7] There has, however, been no
comparison of Shelley's and Byron's writing on this subject, or of the way in

---

[4]   'Ode on Melancholy', 28; 'The Eve of St Agnes', stanza 30; 'Ode to a Nightingale', 16,
and *Hyperion*, I. 189.
[5]   Georg Wilhelm Friedrich Hegel, *On Art, Religion, Philosophy, Introductory Lectures to
the Realm of Absolute Spirit*, ed by J. Glenn Gray (New York: Harper and Row, 1970) p. 67.
[6]   The *Oxford English Dictionary* does not record any use of the word 'vegetarian' before
1839 and the term does not seem to have been in general use until the formation of the
Vegetarian Society in England in 1847. The terms commonly used in the Romantic period
were 'Pythagorean' or 'Brahmin'. However, Thomas comments (p. 295) that 'By the
beginning of the eighteenth century ... all the arguments which were to sustain modern
vegetarianism were in circulation.' The word 'vegetarian' is said by Tester (p. 143) to be
cognate not with 'vegetable' but with the Latin *vegetus,* meaning 'whole, sound, fresh and
lively'.
[7]   Timothy Morton, *Shelley and the Revolution in Taste: the Body and the Natural World*
(Cambridge: Cambridge University Press, 1994) p. 2 and *passim;* Peter W. Graham, 'The
Order and Disorder of Eating in Byron's *Don Juan*' in *Disorderly Eaters: Texts in Self-
Empowerment,* ed by Lilian J. Furst and Peter W. Graham (Pennsylvania: University of
Pennsylvania Press, 1992) pp. 113–23; Wilma Paterson, *Lord Byron's Relish: The Regency
Cookery Book* (Glasgow: Dog and Bone, 1990); Jeremy Hugh Barron, 'Illness and Creativity:
Byron's Appetites, James Joyce's Gut, and Melba's Meals and Mésalliances,' and Arthur
Crisp, 'Ambivalence towards Fatness and its Origins' in *BMJ* 7123 (20–27 December 1997)
pp. 1697–703.

which their attitudes to animals are specifically connected with their presentation of ideas about food.

Anthropological approaches

Shelley perceived the choice and consumption of food as political activities, highly conditioned by the culture in which they are located; and a commentary upon food and eating forms an important part of Byron's self-presentation as a cosmopolitan and detached – effectively, an anthropological – observer of contrasting cultures. 'L'univers est une espèce de livre, dont on n'a lu que la première page quand on n'a vu que son pays' ('The universe is a kind of book, of which one has only read the first page when one has seen only one's own country'): the sentence from Fougeret de Monbron's *Le Cosmopolite, ou le Citoyen du Monde* which Byron chose as the epigraph for the first two cantos of *Childe Harold* reflects the extent to which, as a 'citizen of the world' himself, Byron considered himself as increasingly able to lay aside national prejudices in order to appreciate and understand patterns of behaviour different from those in which he was raised. Reflecting the poets' chosen approaches to the topic, this chapter therefore also applies structural anthropological approaches to Romantic-period writing in this area, using Claude Lévi-Strauss's and Mary Douglas's theories concerning strategies of categorization and taxonomy which separate human beings and animals and distinguish between different kinds of foods.

In the second chapter of *Animals and Society*, Keith Tester describes Lévi-Strauss's famous work on the naming of animals and birds (which groups different types of creatures on the basis of their metaphorical or metonymical resemblance to human beings) as a demonstration of the human propensity to 'proceed by pairs of contrasts': claiming that 'a core of social life is the attempt to establish what it is to be properly human, to establish human uniqueness in contradistinction to the otherness of the natural environment'.[8] Tester presents (pp. 35–6) Lévi-Strauss's understanding of the social restrictions on meat-eating as being closely allied to this:

> Social prohibitions on meat represent an attempt to deny human and animal consubstantiality. Consubstantiality is rooted in the ability of humans to 'assimilate the flesh' of animals, and it is repudiated because, Lévi-Strauss suggests, any overt acceptance of the ability of human bodies to take in the bodies of animals 'would imply recognition on the part of man of their common nature. The meat of any animal species must therefore not be assimilated by any group of men.' In other words, humans perceive the fact that we are like animals in that we are similarly corporeal, but that recognition is the basis of a system of dietary classifications which seek to deny it and, once more, entrench the gulf between humans and animals. Social restrictions develop around meat-eating because, by saying that humans and animals are the same, society is able to demonstrate that they are different.

[8] Tester's references are to Claude Lévi-Strauss, *The Savage Mind*, pp. 204–8.

This defines a sphere which, I suggest, is common – and commonly important – to Shelley's and Byron's operations, although with a marked difference of emphasis. In Shelley's work the reader is mainly conscious of his engagement with the latter part of the formula: the personal and social control which distances human beings from animals, even (or especially) in our eating habits. This is evident in Shelley's prose writings on vegetarianism and in passages like the following one from *Queen Mab,* where, as Morton suggests (p. 87), the Golden Age is seen as one where 'animals forget their animality, and humans become more humane':

> And where the startled wilderness beheld
> A savage conqueror stained with kindred blood,
> A tigress sating with the flesh of lambs
> The unnatural famine of her toothless cubs,
> Whilst shouts and howlings through the desert rang,
> Sloping and smooth the daisy-spangled lawn,
> Offering sweet incense to the sun-rise, smiles
> You see a babe before his mother's door,
>   Sharing his morning's meal
>   With the green and golden basilisk
>   That comes to lick his feet.
>   (VIII. 77–87)

Although animals are to be protected from human cruelty and even called 'brethren' – and Shelley's biographers cite several instances of his attempts to rescue animals from cruel treatment – the very process of showing such humanitarianism can be seen as one which distances the human from other species and, like the Wilberforcean parliamentarians described in Chapter Three, seeks to legislate against cruelty to animals in order to render human beings less like 'the brutes' they might otherwise resemble.[9] Dr William Lambe, whose work and ideas were a major source for Shelley's vegetarian writings, expresses this approach particularly clearly when he writes:

> In his nobler part, his rational soul, man is distinguished from the whole tribe of animals by a boundary, which cannot be passed. It is only when man divests himself of his reason, and debases himself by brutal habits, that he renounces his just rank among created beings and sinks himself below the level of the beasts. *(Additional Reports,* p. 226–7)

This is the characterization of animals that Shelley himself draws on when he describes 'the narrow and malignant passions which have turned man against man as a beast of blood, [and] the enlightened brutality of the multitude'; and it is in this spirit, when he wishes to show human beings in their worst light (as in the final act of *Swellfoot the Tyrant* when Swellfoot and the members of his Court are turned into animals) that Shelley resorts to a shorthand which uses the names of animal species as little more than terms of abuse:

---

[9]   For examples of Shelley's attempts to rescue distressed animals, see Morton, pp. 75–6 and Thomas Jefferson Hogg, *Life of Percy Bysshe Shelley,* 2 vols (London: Edward Moxon, 1858) I. 120–2.

Come, let us hunt these ugly badgers down,
These stinking foxes, these devouring otters,
These hares, these wolves, these anything but men.
(II. ii. 116–18)[10]

By contrast, Byron seems more obviously interested in the consubstantiality or consanguinity of human beings and animals, which he sometimes views with disgust, but more often with relish, or by satirically and thereophilically giving the moral advantage to the animals.

In *Purity and Danger,* Mary Douglas, drawing on Lévi-Strauss's work, considers the 'danger and uncleanness' which attach to objects that cannot be made to fit into the 'hard lines and clear concepts' which humankind's categorizing tendency demands (p. 162). Thus when the books of Deuteronomy and Leviticus lay down dietary laws for the Israelites, she sees them as defining cleanness in animals not through what in modern terms would be identified as the beasts' hygienic or unhygienic qualities, but through the underlying principle of conforming fully to their class. Leviticus, she claims, allots to each element (the earth, the waters and the firmament) its proper kind of animal life, so that

> In the firmament two-legged fowls fly with wings. In the water scaly fish swim with fins. On the earth four-legged animals hop, jump or walk. Any class of creatures that is not equipped for the right kind of locomotion in its element is contrary to holiness. ... Those species are unclean which are imperfect members of their class, or whose class itself confounds the general scheme of the world. (p. 55)

According to this theory, the search for purity is an attempt to force experience into logical categories of non-contradiction. However, as Douglas points out, experience is not fully amenable to this process, and those who make the attempt find themselves led into contradiction. The final part of her argument concerns the danger and power which are attached to deliberate transgression of the boundaries which have been laid down: 'Ritual which can harness these for good is harnessing power indeed', she says (p. 161). In the eating of any given culture,

> There are different kinds of impossibilities, anomalies, bad mixings and abominations. Most of the items receive varying degrees of condemnation and avoidance. Then suddenly we find that one of the most abominable or impossible is singled out and put into a very special kind of ritual frame that marks it off from other experience. The frame ensures that the categories which the normal avoidances sustain are not threatened or affected in any way. Within the ritual frame the abomination is then handled as a source of tremendous power. (p. 167)

Douglas cites (p. 168) the eating of a 'hybrid monster' – the Pangolin or scaly ant-eater – by the Lele people of the Congo, as an example of 'cults which

---

10 From 'An Essay on the Vegetable System of Diet', in *Shelley's Prose, or the Trumpet of a Prophecy,* ed by David Lee Clark (Albuquerque: University of New Mexico Press, 1954) p. 91.

invite their initiates to turn round and confront the categories on which their whole culture has been built up and to recognise them for the fictive, man-made, arbitrary creations that they are'.

Some of Shelley's and Byron's most forceful evocations of iconoclastic eating in *Laon and Cythna, Childe Harold* and *Don Juan* can, I think, be elucidated by the process Douglas outlines here: as a means of 'confronting the categories on which culture has been built up' with an invitation to 'recognise them for the man-made, arbitrary creations that they are'.[11] What fascinates both poets is the *dangerous instability* of the separating, categorizing processes Lévi-Strauss and Douglas describe, and these poems deliberately deploy forms of eating which bring together 'elements which culture keeps apart' in order to create an excitingly powerful charge which is then used to fire an attack on other deep structures and social and political institutions of their time. The 'frame' which 'ensures that the categories which the normal avoidances sustain are not threatened or affected', and which enables the powerful 'abomination' to be handled, is that of a strict literary form with many traditional features. In each case, the poet makes use of an established stanza with a complex, demanding rhyme-scheme, set forth in a long poem which invokes the ancient and respected features of epic, to provide a secure container in which his 'hybrid monsters' may be conjured up and displayed.

Laon and Cythna

My focus for exploring this process in Shelley's work is an examination of the way in which unconventional eating is treated in *Laon and Cythna,* where it is often combined with erotic and sexual motifs or with animal imagery, and sometimes with both, to create these excitingly unstable situations, and to enhance the revolutionary sexuality and politics of the poem. As mentioned above, Shelley's figurative use of animals is often drawn in a rather stereotypical way from the emblematic tradition dating back to the mediaeval period and earlier, whereby different species of animals provide a ready-made set of symbols for human qualities. Thus his dogs are almost always beasts of prey controlled by tyrants (perhaps alluding to Shakespeare's 'dogs of war' in *Julius Caesar);* his tigers are bloodthirsty, fierce and cruel; his snakes usually venomous and sly; his worms corrupting; his birds joyful, and his antelopes innocent, gentle and shy.[12] But *Laon and Cythna* is punctuated by images

---

[11] For references to *Laon and Cythna,* I have used Shelley's original title (under which it was printed but suppressed pending revision by the publishers, C. and J. Ollier, in November 1817) rather than *The Revolt of Islam* (as it was published with revisions in January 1818), because it was specifically the references to incest which were disguised in or removed from the latter version. The text I have quoted is that in *The Complete Poetical Works of Percy Bysshe Shelley,* ed by Neville Rogers, 2 vols (Oxford: Clarendon Press, 1974–75) II. 99–273; since in this version, Rogers claims (p. 361), he has 'preferred what Shelley preferred and restored what he was forced to change'.

[12] *Julius Caesar* III. i. 273. *Prometheus Unbound,* for example, is full of hunting metaphors: from the Furies, which 'as lean dogs pursue / Through wood and lake some struck and sobbing fawn, / Track all things that weep, and bleed, and live' (I. i. 452), to the Chorus's

which combine animals, human beings and iconoclastic eating and sexuality in much more challenging formulations.

As early as the visionary opening sequence, for example, the narrator is disturbed when the 'Woman beautiful as morning' takes the Serpent to her breast:

> And she unveiled her bosom, and the green
> And glancing shadows of the sea did play
> O'er its marmoreal depth: – one moment seen,
> For ere the next, the Serpent did obey
> Her voice, and, coiled in rest in her embrace it lay. ...
>
> I wept. 'Shall this fair woman all alone
> Over the sea with that fierce Serpent go?
> His head is on her heart, and who can know
> How soon he may devour his feeble prey?'
> (I. 20–2 [13])

Shelley's restructuring of the Judaeo-Christian myth transforms the story of the woman and the serpent not only (in later stanzas) into that of actual lovers, but also into this quasi-maternal scene where the proffering of food and the eating of it is reversed from the Eden myth, so that an erotic but also maternal female becomes the potential source of food for an ambiguously dangerous but childlike male. The scene is a locus for Shelley's fascination with the maternal, and the combination of the maternal and the erotic identified by Barbara Charlesworth Gelpi, although Gelpi does not cite this particular example. [14] It also brings together the human and non-human: the way in which the serpent 'obey[s] / Her voice' and 'rest[s] in her embrace' are both childlike and animal-like. This combination is deeply complicated by the narrator's anxiety about devouring and prey, and these and his sexual curiosity about the union of a woman and a phallic animal are signs that he is being faced in his dream with a radical reordering of distinctions and values.

This anxious combination of maternity, eroticism, feeding and animals is a recurrent theme in the poem, and even in an intentionally wholesome scene such as that in Canto V, where one of the sculptures in Laone/Cythna's temple shows:

> A Woman sitting on the sculptured disk
>   Of the broad earth, and feeding with one breast
> A human babe and a young basilisk
> (V. 50),

the emblematic significance of the group ('represent[ing] the reconciliation of humanity with nature', according to Morton) is actually much less striking to

---

'hungry Hours' which, like hounds, 'chased the Day like a bleeding deer, / And it limped and stumbled with many wounds / Through the nightly dells of the desert year' (IV. 73).

[13] References to *Laon and Cythna* are given in the form of canto number followed by stanza number(s).

[14] Barbara Charlesworth Gelpi, *Shelley's Goddess: Maternity, Language, Subjectivity* (New York: Oxford University Press, 1992).

the reader than the repellent image of a reptile being suckled by a woman. Shelley intends this symbol to provide a positive, nurturing celebration of the removal of barriers between human beings and animals, brought about by humankind's renunciation of meat-eating, to which the central image of *Prometheus Unbound* is the negative and destructive opposite. In the drama, again, a human being (Prometheus) feeds an animal from the human's own substance: but the feeding eagle is the instrument of the tyrant Jupiter and, although the substance of the human feeder is – like breast-milk – constantly renewed, the male source of the food experiences anguish and agony from the feeding, instead of joy.[15] Both healthful and unhealthful feeding can be imaged through breast-feeding, as the portrayal of the celebrating Earth demonstrates after Prometheus is freed:

> Henceforth the many children fair
> Folded in my sustaining arms – all plants,
> And creeping forms, and insects rainbow-winged,
> And birds and beasts and fish, and human shapes
> Which drew disease and death from my wan bosom,
> Draining the poison of despair – shall take
> And interchange sweet nutriment.
> (III. iii. 90–6)

Gelpi and Alan Richardson have shown how male Romantic poets attempted to appropriate feminine empathy, sympathy and other qualities within their writing: drawing in particular on 'memories and fantasies of identification with the mother in order to colonize the conventionally feminine domain of sensibility'.[16] They both deploy the image of Shelley's attempts to breast-feed his daughter Ianthe, when Harriet Shelley refused to do so, as an example of Shelley's intense identification with motherhood. But the context of this scene in fact reflects as much Shelley's anxiety about eating in general, and the possibility of the child's imbibing the wet-nurse's soul through the nursing, as it does his wish to mother the child himself: 'at last, in his despair, and thinking that the passion in him would make a miracle, he pulled his shirt away and tried himself to suckle the child.'[17]

---

[15] The bringing together of animal and food imagery in the play is, of course, highly deliberate, and reflects Shelley's (re)deployment of the Prometheus myth in his vegetarian writing: 'Prometheus (who represents the human race) effected some great change in the condition of his nature, and applied fire to culinary purposes; thus inventing an expedient for screening from his disgust the horror of the shambles. From this moment his vitals were devoured by the vulture of disease' (Shelley, *Prose,* ed by Murray, I. 78).

[16] Alan Richardson, 'Romanticism and the Colonization of the Feminine' in *Romanticism and Feminism,* ed by Anne K. Mellor (Bloomington: Indiana University Press, 1988) pp. 13–25 (p. 13).

[17] Newman Ivey White, *Shelley,* 2 vols (New York: Knopf, 1940) I. 326. Shelley may have been recalling a footnote to Canto II. 122 of Erasmus Darwin's poem *The Temple of Nature,* where 'it is affirmed, that some men have given milk to their children in desert countries where the mother has perished; as the male pigeon is said to give a kind of milk from his stomach along with the regurgitated food, to the young doves'. *The Temple of Nature; or, the Origin of Society: A Poem, with Philosophical Notes,* by Erasmus Darwin, ed by E. D. Menston (London: Scolar Press, 1973) p. 53.

In Canto VII of *Laon and Cythna* the anxiety about distinctions between human beings and animals and about eating and being eaten is intensified when it is brought into relation with pregnancy and childbearing itself. In stanza 15, Cythna is imprisoned in a womb-like cave, where she is visited only by a sea-eagle which brings her food, and is agonized by madness which for a while makes 'the sea-eagle look a fiend, who bore/[Laon's] mangled limbs for food'. She feels she is about to become a mother (stanza 16):

> 'Another frenzy came – there seemed a being
>   Within me – a strange load my heart did bear,
> As if some living thing had made its lair
>   Even in the fountains of my life',

and this image of an animal 'lairing' within her is repeated at stanza 25, after her baby has been born and then taken away:

> 'So, now my reason was restored to me
>   I struggled with that dream, which, like a beast
> Most fierce and beauteous, in my memory
>   Had made its lair, and on my heart did feast.'

The second beast is ambivalently characterized as both 'fierce' and 'beauteous', but the fact that it feasts on a human heart makes it more like Prometheus's feeding eagle than the basilisk suckled by the woman in Cythna's temple. The representation of both the baby and its dream-counterpart as a beast, living and eating inside a human body, expresses powerful fears about the experience of pregnancy and childbearing which do not accord with Gelpi's picture of Shelley's envy of the fascination and mystique of motherhood, or with Richardson's vision (p. 20) of male Romantic poets' goal of 'constant ... appropriation of feminine, maternal characteristics and functions'. Shelley's fantasy seems to have more in common with themes of grotesque bodily invasion which give rise to monstrous births, featured in written science fiction such as John Wyndham's *The Midwich Cuckoos* and films like *The Invasion of the Body Snatchers* and *Alien,* where the birth of the alien creature from a human 'mother' (who may be of either sex) is represented as a form of evil cloning. If, as Gilbert and Gubar propose, Mary Shelley intends to show in Victor Frankenstein the disastrous results of a male taking upon himself the woman's role of childbearing, Percy Shelley's attempt to imaginatively enter into this role in *Laon and Cythna* also results in imagery which reveals, not envy of a beneficent function, but instead a complex revulsion.[18]

Laon and Cythna renders eating aberrant not only by incongruous combinations with animals, childbirth and breast-feeding, but also by linking it with activities which deliberately challenge social structures and transgress established boundaries. Thus in Canto III Shelley makes the starving Laon dream or imagine that he has eaten his dead sister's body, in an act which

---

[18] Sandra M. Gilbert and Susan Gubar, *The Madwoman in the Attic: The Woman Writer and the Nineteenth-century Literary Imagination* (New Haven: Yale University Press, 1979) p. 232.

combines abhorrent eating of rotten flesh not only with cannibalism but also
with incest:

> eagerly, out in the giddy air,
> Leaning that I might eat, I stretched and clung
> Over the shapeless depth in which those corpses hung.
>
> A woman's shape, now lank and cold and blue,
>   The dwelling of the many-coloured worm,
> Hung there; the white and hollow cheek I drew
>   To my dry lips – what radiance did inform
>   Those horny eyes? whose was that withered form?
> Alas, alas! it seemed that Cythna's ghost
>   Laughed in those looks, and that the flesh was warm
> Within my teeth!
> (III. 25–6).

The way in which the 'giddy air' and 'shapeless depth' are brought together
with the 'flesh ... warm / Within my teeth' suggests that this passage might be
read alongside Shelley's 'Marianne's Dream' (1817) as evidence of the interest
he shared with Leigh and Marianne Hunt in the way physical factors such as
drinking alcohol and over-eating affect dreams and create nightmares. Leigh
Hunt wrote facetiously in 1820:

> It is certain enough, however, that dreams in general proceed from
> indigestion. ... The inspirations of veal, in particular, are accounted
> extremely Delphic; Italian pickles partake of the same spirit of Dante;
> and a butter-boat shall contain as many ghosts as Charon's.[19]

While elsewhere Shelley too treats the visions arising from indigestion with
a mixture of flippancy and terror, this passage does perform more than a
Gothic fascination with the macabre: aiming, within the revolutionary setting
of the poem, to raise the temperature of the reader's emotions into anger
against tyrannical torture. As Shelley claims in the Preface to the poem, the
incest in the original version was 'intended to startle the reader from the trance
of ordinary life' and to 'strengthen the moral sense by forbidding it to waste its
energies in seeking to avoid actions which are only crimes of convention'; and
the purpose of passages such as this is to open horizons beyond institutionally
approved emotions and to invoke a 'hybrid monster' as a way of confronting
and exposing the cultural categories which Shelley wanted to reshape and
reform.
    In Canto X, however, when Laon and Cythna's revolution has been
defeated, Shelley asks us to accept a negative reading of the coming together
of creatures which are normally kept apart: seeing it as an effect of the collapse
of the restraining power of human civilization, rather than a positive and

---

[19] *Essays by Leigh Hunt,* ed by Arthur Symons (London: Walter Scott, 1888) p. 237. See
Timothy Clark and Mark Allen, 'Between Flippancy and Terror: Shelley's "Marianne's
Dream"' in *Romanticism* I.1 (1995) pp. 90–105. Byron accused Keats of 'viciously soliciting
his own ideas into a state which is neither poetry nor any thing else but a Bedlam vision
produced by raw pork and opium' (letter to John Murray, 9 November 1820: *BLJ* VII. 225).

healthful breaking down of barriers. As war, famine, pestilence and starvation take over the land, creatures which are naturally antipathetic join together to feed: both wild animal ones:

> then meet
> The vulture, and the wild dog, and the snake,
>   The wolf, and the hyena gray, and eat
> The dead in horrid truce
> (X. 3);

and human ones: the Kings and Priests, who

> swore
> Like wolves and serpents to their mutual wars
> Strange truce, with many a rite which Earth and Heaven abhors.
> (X. 7)

The poem demonstrates how the human need for 'hard lines and clear concepts' will attempt to assert itself even in situations where dissolution and entropy threatens the entire system of cultural categories: the cannibalism of Canto X is the more horrific because it is not wild or savage, but a deliberate part of a human system which still controls the minutiae of life:

> There was no corn – in the wide market-place
>   All loathliest things, even human flesh, was sold;
> They weighed it in small scales.
> (X. 19)

Motherhood is, again, characterized as a source of anguish when brought into conjunction with feeding and eating:

> The mother brought her eldest-born, controlled
> By instinct blind as love, but turned again
> And bade her infant suck, and died in silent pain.
> (X. 19)

Finally, while the plague rages, the complete disintegration of the social system is demonstrated by the way in which the most basic distinction – that between the self and the other – is lost:

> Many saw
> Their own lean image everywhere, it went
> A ghastlier self beside them, till the awe
>   Of that dread sight to self-destruction sent
> Those shrieking victims.
> (X. 22)

It is only through this total loss of civilization and the breakdown of the most fundamental categories and structures that Shelley's vision of a new, post-revolutionary world can be brought into being. The Wild West Wind of the 'Ode' must be 'destroyer' before it can be 'preserver', it must be 'uncontrollable', and 'tameless, and swift and proud' in its destruction of outworn systems, before the new seeds and sparks, scattered through the earth by its activity, can be brought to life.[20] Catastrophic destruction and recreation of the whole earth, as envisaged in the geological theories of Baron Cuvier and James Parkinson and invoked in Act IV (296–318) of *Prometheus Unbound*, may even be required to bring about real change in humanity's condition. The 'story' of *Laon and Cythna* was, Shelley claims in the Preface to the poem, chosen 'in contempt of all artificial notions or institutions'; its object is 'to break through the crust of those outworn opinions on which established institutions depend', and the way in which the poem is to recommend 'motives which I would substitute for those at present governing mankind' is by 'awaken[ing] the feelings' of its readers rather than by 'methodical and systematic argument'. The passages which have been considered here show that the statement of Shelley's revolutionary ideology, and those aspects of it which are concerned with animals, food and eating, are indeed not restricted to the overtly political expressions of the poem. They are not confined to the obviously positive and uplifting images which occur at the points where the hero and heroine and their allies deliberately state their beliefs, nor to the points where these beliefs are put into practice in scenes such as the revolutionary vegetarian feast in Canto V. Shelley's politics of consumption are also expressed at a deeper, 'feeling', level by imagery which activates the excitement and energy caused by challenging and breaking the categories and taboos with which human beings surround their food and eating habits, and which they use to maintain careful distinctions between themselves and other animals.

Shelley's iconoclasm in matters concerning food and eating is, both here and in his vegetarian prose, overt and intentionally shocking. In *A Vindication of Natural Diet* (1813), for example, he brings an element of deliberate drama into his paraphrase of Plutarch's essay 'On the Eating of Flesh' from the *Moralia* XII, in order to highlight the unacceptably fragile distinction between live and dead flesh. Plutarch's text argues that human beings are not anatomically equipped to eat meat, by describing how difficult it would be for a human to kill an animal bare-handed:

> just as wolves and bears and lions themselves slay what they eat, so you are to fell an ox with your fangs or a bear with your jaws, or tear a lamb or hare in bits. Fall upon it and eat it still living, as animals do.[21]

Shelley's account reads:

> Let the advocate of animal food force himself to a decisive experiment on its fitness, and, as Plutarch recommends, tear a living lamb with his teeth, and *plunging his head into its vitals* slake his thirst with the steaming blood. (Shelley, *Prose*, I. 80; my italics)

---

[20] See William D. Brewer, *The Shelley-Byron Conversation* (Florida: University Press, 1994) p. 33. Morton comments of the 'Ode' (p. 214): 'nature could be seen not as timeless, but as a dynamic movement of "disorganization and reproduction". ... Nature, like the social order of which it is an analogue, may be conceived as a machine which "works" better through breakdown.'

[21.] 'On the Eating of Flesh' in *Plutarch's Moralia*, XII. 553.

Shelley's version induces far greater anxiety in the reader than Plutarch's does because – by evoking the possibility of placing, or actually 'plunging' one's head into the carcass – it illustrates vividly how insecure the distinction is between body and food, eating and being eaten, and in particular how human and animal substance are confounded together in the act of eating meat, and thus calls up the unacceptable 'consubstantiality' of which Lévi-Strauss writes. The kind of thinking on which Shelley was drawing is evidenced by John Oswald's contention in *The Cry of Nature* (1791, pp. 17–18) that 'animal food overpowers the stomach, clogs the functions of the soul, and renders the mind material and gross. In the difficult, the unnatural task of converting into living juice the cadaverous oppression, a great deal of time is consumed, a great deal of danger is incurred.' Shelley sums up his vegetarian philosophy, at the end of the note to *Queen Mab*, in similar terms: 'NEVER TAKE ANY SUBSTANCE INTO THE STOMACH THAT ONCE HAD LIFE.' Seen from the human point of view, eating means that the human substance engulfs the animal, but here the act of incorporation is rendered horrific by reversing the process, making the animal's body take in the human head.

Sceptical carnivorousness

Byron's approach to themes of iconoclastic eating, by contrast, is more apparently conciliatory towards the reader and less overtly politically confrontational, and the connection between radical politics and abstention from meat which Shelley evokes is one of the many dietary systems subjected to Byronic scepticism and irony. Byron was fascinated by the role of eating in religious practice, and by the way religious ritual seems arbitrarily to forbid some foods while authorizing other forms of apparently unspeakable eating. Thus in April 1814, he described to Thomas Moore how he 'nearly killed [him]self' eating a collar of brawn, and tells Moore, who was a Roman Catholic, 'All this gourmandize was in honour of Lent; for I am forbidden meat all the rest of the year – but it is strictly enjoined me during your solemn fast' *(BLJ,* IV. 92). And in March 1822, in recounting to Moore how he was 'breeding one of [his] daughters a Catholic,' he characterizes Roman Catholicism as

> by far the most elegant worship, hardly excepting the Greek mythology. What with incense, pictures, statues, altars, shrines, relics, and the real presence, confession, absolution, – there is something sensible to grasp at. Besides, it leaves no possibility of doubt; for those who swallow their Deity, really and truly, in transubstantiation, can hardly find anything else otherwise than easy of digestion. *(BLJ,* IX. 123)

Two scenes – one the 'caritas romana' vision from *Childe Harold* Canto IV, and the other the shipwreck scene from *Don Juan* Canto II – can be used to show how Byron utilizes this awareness in order to perform a process similar to that which I have identified in *Laon and Cythna,* in terms that Mary Douglas describes as 'confronting the categories on which culture has been built up' with an invitation to 'recognise them for the man-made, arbitrary creations that

they are'. The 'caritas romana' scene in stanzas 148 to 151 of *Childe Harold
IV* is one of a series in Byron's work which use eating and captive animals as a
way of exploring the experience of imprisonment.[22] So Tasso in the 'Lament'
eats 'tasteless food', 'alone' and learns to

> banquet like a beast of prey,
> Sullen and lonely, crouching in the cave
> Which is my lair, and – it may be – my grave.
> ('The Lament of Tasso', 15–17)

The food of the brothers imprisoned in the fortress of Chillon is a token of
their imbrutement by a tyrannical political regime:

> Our bread was such as captives' tears
> Have moisten'd many a thousand years,
> Since man first pent his fellow men
> Like brutes within an iron den.
> ('The Prisoner of Chillon', 134–7)

The 'caritas romana' scene opens in 'a dungeon', 'dim' and 'drear', and
uses a vision of iconoclastic feeding, in which a young mother suckles her own
father, to enact a scene in which 'sacred nature's decree' is 'reversed' in order
to arrive at a celebration of the replenishing of 'life' – 'as our freed souls rejoin
the universe'.

> There is a dungeon, in whose dim drear light
> What do I gaze on? Nothing: Look again!
> Two forms are slowly shadowed on my sight –
> Two insulated phantoms of the brain:
> It is not so; I see them full and plain –
> An old man, and a female young and fair,
> Fresh as a nursing mother, in whose vein
> The blood is nectar: but what doth she there,
> With her unmantled neck, and bosom white and bare?

> Full swells the deep pure fountain of young life,
> Where *on* the heart and *from* the heart we took
> Our first and sweetest nurture, when the wife,
> Blest into mother, in the innocent look,
> Or even the piping cry of lips that brook
> No pain and small suspense, a joy perceives
> Man knows not, when from out its cradled nook
> She sees her little bud put forth its leaves –
> What may the fruit be yet? – I know not – Cain was Eve's.[23]

---

[22] Byron seems to allude to Rousseau's association of animals and liberty *(Discourses and
Social Contract,* p. 93): 'when I see free-born animals dash their brains out against the bars of
their cage, from an innate impatience with captivity; when I behold numbers of naked savages,
that despise European pleasures, braving hunger, fire, the sword, and death, to preserve
nothing but their independence, I feel it is not for slaves to argue about liberty'.
[23] Erasmus Darwin described in *Zoonomia, or, the Laws of Organic Life* (2 vols, London:
J. Johnson, 1796, I. 126) how 'the females of lactiferous animals have another natural inlet of

But here youth offers to old age the food,
The milk of its own gift: it is her sire
To whom she renders back the debt of blood
Born with her birth. No; he shall not expire
While in those warm and lovely veins the fire
Of health and holy feeling can provide
Great Nature's Nile, whose deep stream rises higher
Than Egypt's river: – from that gentle side
Drink, drink and live, old man! Heaven's realm holds no such tide.

The starry fable of the milky way
Has not thy story's purity; it is
A constellation of a sweeter ray,
And sacred nature triumphs more in this
Reverse of her decree, than in the abyss
Where sparkle distant worlds: – Oh, holiest nurse!
No drop of that pure stream its way shall miss
To thy sire's heart, replenishing its source
With life, as our freed souls rejoin the universe.

Byron's presentation of the legend is authorized by ancient chroniclers and supported by Hobhouse's studious notes both alongside the publication of the verse and separately in the *Historical Illustrations,* and these help to provide the ritual 'frame' which Douglas describes as necessary for the safe handling of the powerful abomination. But, as we might guess from the reference to Cain (a revolutionary, vegetarian eater, amongst other things, as Byron's eponymous play makes very clear), what Byron actually leads us into viewing is a scene where the feeding and being fed is profoundly unorthodox: an erotic and incestuous incident with overtones of cannibalism supposedly made respectable by the authority of the ancient Roman tale, the invocation of innocent motherhood and the classical fable of Hera, Heracles and the Milky Way. The aberrant eating of this passage thus cites it firmly within the iconoclastic, anti-fabulous tone, with the abhorrence of the loss of liberty through imprisonment and the strong message of antipathy to the restoration of the old order in post-Napoleonic Europe, which is a project of this Canto as a whole.

The shipwreck scene in *Don Juan* Canto II, lines 329–880 presents a microcosm of society *in extremis,* where entropy threatens the 'hard lines and clear concepts' of behaviour which humankind craves, with a resulting powerful mingling of ritual and barbarism. Unlike the treatment of the theme by Gothic predecessors such as Matthew Lewis in *The Monk,* or Shelley's dreamlike sequences, Byron goes out of his way to present cannibalism as something which is 'real' and could arise in ordinary life. As Peter Graham points out (pp. 115–16), the approach to the cannibalistic scene is 'slow and detail-laden': careful ceremony is followed in the drawing of lots to see who should perish to feed the others (II. 589–600); Pedrillo requests to be allowed to die a Catholic (II. 601–8), and his death parodies the central Christian scene

---

pleasure or pain from the suckling of their offspring'. Richardson (p. 17) cites this passage as one in which Byron usurps a feminine role by 'develop[ing] a figure [which] ... combines soul-feeding with breast-feeding'.

of the crucifixion, with an innocent victim willingly laying down his life to save his friends. However, those who eat Pedrillo's flesh are not saved by this sacramental meal but on the contrary die blaspheming, foaming and rolling, drinking salt water, grinning, howling, screeching, swearing, and laughing like hyenas (II. 625–32). Juan survives, not necessarily because he is resolute or brave, but because he successfully negotiates the intricacies of the ordeal by food: he does not eat when culturally he should not, and is practical about doing so when he should. Although, like Lévi-Strauss, Byron is highly conscious of European taboos about eating dogs – especially pet ones which share human social life and are 'metonymical to human beings' – he allows Juan eventually to succumb, like Byron's own shipwrecked grandfather, to eating the flesh of his pet in order to stay alive.[24]

Graham believes that stanza 67 of Canto II, which occurs at a crucial point in the shipwreck scene, and which seems to confirm Byron's approval of human carnivorousness, is one of the places where he gives readers 'every assurance that these transgressors of our eating codes have done all that they can to avoid what they end up doing' (p. 116):

> But man is a carnivorous production,
>   And must have meals, at least one meal a day;
> He cannot live, like woodcocks, upon suction,
>   But, like the shark and tiger, must have prey:
> Although his anatomical construction
>   Bears vegetables in a grumbling way,
> Your labouring people think, beyond all question,
> Beef, veal and mutton, better for digestion.
>
> And so it was with this our hapless crew . . .
> (II. 529–37)

However, I suggest that the assurance of a clear moral stance here is deceptive, and many of the clauses which seem most definite are in fact tentative or can be read as deliberately provocative about meat-eating and vegetarianism. 'Man is a carnivorous production', in the first line, for example, is qualified and requalified in the second line by replacing the expected 'meat' twice with the weaker 'meals': man must indeed have 'meals', but need they be 'meat' ones, Byron implies. In the fourth line we are told that human beings 'must have prey', but the sanguinary connotations of 'prey' – again a substitute for 'meat' – warn us that this is not a simple and straightforward statement of fact, but a particular representation of humanity, which is not the way it 'must' be at all. We are led to consider that, if it is natural for humankind to have 'prey', whether it is also natural for them to prey on their own kind, and this can be interpreted either as eating or as making war upon their fellows. Byron had meditated upon these issues at greater length in *The*

---

[24] *The Savage Mind*, p. 205. Byron refers to *A Narrative of the Honourable John Byron* (1768) which gives an account of the hardships suffered by Admiral 'Foulweather Jack' Byron following the wreck of the *Wager* in 1740: see *LBCPW* V. 689. Byron is perhaps deliberately playing with his reputation as misanthropic dog-lover in this passage.

Siege of Corinth, when at a key point Alp's eye pans away from the human action to focus on animals:

> He saw the lean dogs beneath the wall
> Hold o'er the dead their carnival,
> Gorging and growling o'er carcass and limb,
> They were too busy to bark at him!
> From a Tartar's skull they had stripped the flesh,
> As ye peel a fig when its fruit is fresh;
> And their white tusks crunched o'er the whiter skull,
> As it slipped through their jaws, when their edge grew dull,
> As they lazily mumbled the bones of the dead,
> When they scarce could rise from the spot where they fed;
> So well had they broken a lingering fast
> With those who had fallen for that night's repast.
> *(The Siege of Corinth,* 409–20)[25]

The lines which follow make it clear, however, that the fact that beasts and birds view human bodies simply as part of the food chain demeans not them, but the men who have provided such a feast of bodies:

> But when all is past, it is humbling to tread
> O'er the weltering field of the tombless dead,
> And see worms of the earth and fowls of the air,
>   Beasts of the forest all gathering there;
>   All regarding man as their prey,
>   All rejoicing in his decay.
> *(The Siege of Corinth,* 444–9)

Text and context

These lines and the stanza about man's carnivorousness from *Don Juan* can be elucidated by some of the discourses connecting eating, meat and animals which were current in Byron's and Shelley's time. Timothy Morton has demonstrated that Shelley knew well, and drew upon in his own writing the work of a wide range authors in this area, including Pythagoras as presented by Ovid in *Metamorphoses* XV; Plutarch; the seventeenth-century vegetarian Thomas Tryon; Buffon; Rousseau's *Discourse on the Origins and Foundations of Inequality among Men* (1755); Dr William Lambe; Lord Monboddo; Joseph Ritson's *Essay on Abstinence from Animal Food* (1802), and John Newton's *The Return to Nature; or, A Defence of the Vegetable Regimen* (1811).[26]

---

[25] In *Beppo* (line 42) Byron points out that 'carnival' means 'farewell to flesh'.

[26] Ovid, *Metamorphoses,* trans, by A. D. Melville (Oxford: Oxford University Press, 1986); Thomas Tryon, *Pythagoras His Mystic Philosophy Revived* (London: Thomas Salisbury, 1691); Rousseau, *Discourse on the Origins and Foundations of Inequality among Men* (Harmondsworth: Penguin, 1984); James Burnet, Lord Monboddo, *The Origin and Progress of Language* (Edinburgh: J. Balfour and T. Cadell, 1774) and *Antient Metaphysics, Volume Third, Containing the History and Philosophy of Men* (London: T. Cadell, 1784);

Byron may not have known all these, but he was familiar with Shelley's own vegetarian writings, and he also knew at least Monboddo's theories (which he satirizes in an early poem), the works of Buffon and Rousseau, which he often quotes, probably Newton's essay, and Thomas Love Peacock's novel *Melincourt* which contains satire on several aspects of the debate.[27]

The debate to which Byron refers in the first line of the *Don Juan* stanza, about whether human beings are 'naturally' carnivorous, herbivorous or frugivorous, and in the fourth, as to whether Man is a beast of prey, was a recognized part of the vegetarian tradition by this date. Rousseau, for instance, contended that:

> Animals that live only on vegetation all have blunt teeth, like the horse, the ox, the sheep, the hare, while voracious animals have sharp teeth, like the cat, the dog, the wolf, the fox. As for the intestines, frugivorous animals have some, such as the colon, which are not found in voracious animals. It appears therefore that man, having teeth and intestines like the frugivorous animals, should naturally be classed in that category. *(Discourse on ... Inequality,* p. 143)

In Britain Lord Monboddo was famous, or notorious, for claiming that orang-utans (which were thought to live exclusively on fruit) were actually primitive men – 'a barbarous nation which has not yet learned the use of speech' – and this led him to conclude that 'by nature, and in his original state, [Man] is a frugivorous animal, and that he only becomes an animal of prey by acquired habit' *(Origin and Progress of Language,* pp. 270 and 224). Monboddo recounts examples of natural herbivores being taught to live, unnaturally, upon meat – and becoming human-like in their other recreations:

> there was a sheep in my neighbourhood in the country, who, being brought up in the house by hand, learned to eat flesh, and even the flesh of its own species, and became so fond of tobacco, that, after he was restored to his natural life with the flock, he would come up to a gentleman in the field, and eat a piece of tobacco with him. *(Antient Metaphysics,* p. 176)

Monboddo believed that the same kind of degeneration had occurred in human beings, whose nature 'is so pliable that it can suit itself to that [ie meat]

Joseph Ritson, *An Essay on Abstinence from Animal Food, as a Moral Duty* (London: Richard Phillips, 1802); and John Frank Newton, 'The Return to Nature; or, A Defence of the Vegetable Regimen', cited here from *The Pamphleteer,* XIX no. xxxviii (1821) pp. 497–530; XX no. xxxix (1822) pp. 97–118, and XX no. xl (1822) pp. 411–29.

[27] Shelley refers to Pythagoras, Plutarch, Tryon, Ritson and Newton in his *A Vindication of Natural Diet* (Note 17 to *Queen Mab,* 1813) and *On the Vegetable System of Diet* (1814–15) and he ordered a copy of Monboddo's *Origin and Progress of Language* when he was writing *Queen Mab*. For Byron on Monboddo, see *The Edinburgh Ladies' Petition to Doctor Moyes, and his Reply* (1807). Newton's essay of 1811 was reprinted in *The Pamphleteer* in 1821 in numbers which also contained 'A Letter of Expostulation to Lord Byron on his present Pursuits', and instalments of the Pope-Bowles controversy in which Byron was involved. Shelley wrote to Peacock in August 1821 that Byron 'is a great admirer' of *Melincourt*.

or even a worse diet' *(Antient Metaphysics,* p. 176), and that the degeneration into meat-eating led to even more degraded habits:

> I am persuaded, that most nations, at some time or another, have been cannibals; and that men, as soon as they became animals of prey, which, as I have said, they were not originally, fed upon those of their own kind as well as those of other animals. *(Origin,* p. 227)

Joseph Ritson, whose *Essay on Abstinence from Animal Food* was a major source for Shelley's vegetarian writing, similarly claimed that meat-eating would lead to cannibalism, and that 'Those accustom'd to eat the brute would not long abstain from the man', especially since 'when toasted or broil'd,' both would taste much the same (p. 124). Byron's pronouncement that 'Man is a carnivorous production' and 'must have prey' ought to alert us here, therefore, that the shift from 'meat' to 'prey' will almost inevitably lead to 'cannibalism' – and, in the poem, this is what does indeed follow.

Monboddo relates travellers' and naturalists' stories of not only 'orang outangs', but also 'pongos', 'enjockos', 'quimpezes' and other creatures, which were said to show human behaviour by making their bed in a box, hiding their private parts out of modesty, doing 'everything that they are desired to do, by signs or words', and even playing on the pipe, harp and other instruments. One of the proofs, for Monboddo, that orang-utans are the same as human beings is – ironically – their excellent table manners, and he learnt from Buffon of one who 'when he was set at table ... behaved, in every respect, like a man, not only doing what he was bid, but often acting voluntarily, and without being desired' *(Origin,* p. 280).

Monboddo's work draws heavily upon Rousseau, but pushes the argument much further by tenaciously maintaining that social and behavioural as well as anatomical issues must be taken into account in classifying orang-utans as human beings. Monboddo also insists that there is no difference *'now'* between human beings and animals: 'my state of nature is not an imaginary state ... but a real state, upon which we may safely found our philosophy of Man' *(Antient Metaphysics,* p. 68). This was one of the features that led to his work being widely satirized. The *Encyclopaedia Britannica* (ninth edition) states that his writings 'afforded endless matter for jest by the wags of the day', of whom the young Byron was one, and Peacock – rather belatedly – another.[28]

Monboddo's work provided the latter, in *Melincourt,* with the character of Sir Oran Haut-ton: an orang-utan who is also a non-speaking, but gentle, polite, flute-playing and brave baronet, who is to be elected as Member of Parliament for the rotten borough of One-vote. There is a telling touch early in the novel when Sir Oran comports himself perfectly during a dinner-party, helping the guests with great dexterity, and 'show[s] great proficiency in the dissection of game' (p. 42).[29] Peacock's Mr Forester, Sir Oran's friend and minder, has distinctly Shelleyan features, including a conviction that 'the history of the world abounds with sudden and extraordinary revolutions in the

---

[28] *Encyclopaedia Britannica,* ninth edition (London, A. & C. Black, 1881–88) entry for 'Monboddo'.

[29] Thomas Love Peacock, *Melincourt,* ed by Richard Garnett (London: Dent, 1891) p. 42.

opinions of mankind, which have been effected by single enthusiasts' (p. 48), and plans for organizing an 'anti-saccharine festival' in order to demonstrate that 'the use of sugar is economically superfluous, physically pernicious, morally atrocious and politically abominable' (p. 53). Mr Forester's attack on sugar arises primarily, of course, from its being a product of slavery, but his arguments demonstrate how the politically undesirable aspects of this foodstuff are also displaced into other areas, including the dietetic, in a way which resembles the arguments about the eating of meat. The 'blackness' of the food may not be visible to the consumer, but renders it secretly pernicious, recalling Cowper's epigram about the slave-trade, printed in the *Northampton Mercury* in 1792, which also links food with animals and slaves as the objects of cruelty and of protective campaigns in this period:

> To purify their wine some people bleed
> A *lamb* into the barrel, and succeed;
> No nostrum, planters say, is half so good,
> To make fine sugar, as a negro's blood.
> Now lambs and negroes both are harmless things,
> And thence perhaps this wond'rous virtue springs,
> 'Tis in the blood of innocence alone –
> Good cause why planters never try *their own*.[30]

These are issues that Byron touches upon in Canto V of *Don Juan*, when the merchant goes off to dinner after selling Juan and Johnson as slaves – 'that most superior yoke of human cattle'. 'I wonder if his appetite was good?' the narrator asks,

> Or, if it were, if also his digestion?
>   Methinks at meals some odd thoughts might intrude,
>   And conscience ask a curious sort of question,
>   About the right divine how far we should
>   Sell flesh and blood. When dinner has opprest one,
>   I think it is perhaps the gloomiest hour
>   Which turns up out of the sad twenty-four.
>   (V. 233–40)

The 'flesh and blood' of the sixth line is of course deliberately ambiguous: the sale of these commodities can refer both to the animal which one might consume for one's dinner and also to the slave who has been bought and sold to enable one to do so.

Byron is, however, aware of the possibility of another point of view about the undesirability of meat-eating, to which he reverts in the last four lines of the *Don Juan* stanza, mentioned on page 124, which discusses humankind's carnivorous nature. Here he introduces the 'labouring people' who 'think, beyond all question, / Beef, veal and mutton, better for digestion.' In the lean years after Waterloo, real famine threatened the British working classes with a shortage not just of meat but even of the necessities of life, and the

---

[30] The use of animal blood as a purification agent in wine-making continued in Europe until 1997, when it was banned because of the BSE scare. See 'French wine scare over cattle blood' by Susannah Herbert, *Daily Telegraph*, 25 June, 1999, p. 17.

'grumbling' with which labourers and their digestive systems might bear a vegetable diet erupted in 1819 – the year this stanza was published – into open revolt at Peterloo. As William Cobbett commented: 'Meat in the House is a great source of harmony, a great preventer of the temptation to commit those things, which, from small beginnings, lead, finally, to the most fatal and atrocious results.'[31] Meat was perceived as a highly nutritious food, laden with value, not only dietetically but also in terms of class. Keith Thomas (p. 26) quotes the 1748 comment of a Swedish visitor, Pehr Kalm: 'I do not believe that any Englishman *who is his own master* has ever eaten a dinner without meat' (my italics). Meat was also the site, as discussed in Chapter Three, of a massive patriotic war-time investment in 'the roast beef of old England' as the food of British heroes.[32] Even Byron's Juan (who despite being, like the narrator of the poem, ostensibly Spanish) has tastes in food which are very definitely British, and when he awakes on the beach on Haidée's island his first longing is for a beef-steak – although none is forthcoming because, as the narrator punningly tells us, 'beef is rare within these oxless isles' (II. 1225).

This leads to a facetious disquisition on whether 'Pasiphae promoted breeding cattle, /To make the Cretans bloodier in battle' (II. 1239–40), with a reference to the notion, discussed in Chapter Three, that it is by eating beef that 'English people' become 'very fond of war' (II. 1241–5). It is hard to tell how seriously Byron took this version of the ancient doctrine of signatures, whereby plants and animals were thought to demonstrate their use for human beings by the unrelated objects they resembled. He may well have been 'humbugging' or 'mystifying' Lady Blessington, for example, when he told her that 'animal food engenders the appetite of the animal fed upon' (Blessington, p. 86), or when he asked Thomas Moore 'in a grave tone of inquiry', '"Moore, don't you find eating beef-steak makes you ferocious?"' (Moore, *Letters and Journals,* I. 324).[33] But like the universal Englishman nicknamed 'John Bull', who appears frequently in James Gillray's cartoons of this period stuffing down huge amounts of flesh, Byron's presentation of British eating habits demonstrates the force of the proverb 'You are what you eat'.[34]

As Morton (p. 21) comments of Shelley's vegetarianism, 'While meat was too politically charged for Shelley to eat, it was too politically charged for the

[31] William Cobbett, *Cottage Economy* (London: C. Clement, 1822) p. 162.

[32] John Oswald (*The Cry of Nature,* p. 15) called England 'that most carnivorous of all countries', and Hobhouse, visiting Europe just before Waterloo (2 April 1815) noted that 'The English are well received ... generally every where except in France, where they are pursued with cries of *Roast Beef* and other intolerable exclamations'. *Byron's Bulldog: The Letters of John Cam Hobhouse to Lord Byron,* ed by Peter W Graham (Columbus: Ohio State University Press, 1984) p. 187.

[33] Keith Thomas records (p. 291) that Thomas Tryon 'rejected flesh-eating partly because he thought it introduced an animal element into the body, giving man a "wolfish, doggish nature"'.

[34] See, for example, Gillray's print, 'French Liberty and British Slavery' (1792) and, earlier in the century, Hogarth's 'The Roast Beef of Old England'.

poorer classes *not* to eat.'[35] Shelley's political mock-heroic drama *Swellfoot the Tyrant* satirizes this situation, with a hugely fat King and his priests and ministers worshipping at the temple of the goddess Famine, surrounded by starving subjects. However, Shelley's Chorus of 'the Multitude' (following Burke's characterization of 'a Swinish multitude' in his *Reflections on the Revolution in France,* and the New Testament parable of the Gadarene swine) is composed of pigs, and they are portrayed for the most part as foolish, ignorant and gullible, although with real grievances about their lack of food.[36] The figure of Liberty states that even food can be dangerous if too much of it, or the wrong sort, is given to the 'wrong' people:

> I charge thee when thou wake the multitude,
> Thou lead them not upon the paths of blood.
> The earth did never mean her foison
> For those who crown life's cup with poison
> Of fanatic rage and meaningless revenge –
> But for those radiant spirits, who are still
> The standard-bearers in the van of Change.
> (II. ii. 90–6)

Byron in Greece castigated his valet for his 'perpetual lamentations after beef & beer' *(BLJ,* II. 34) and 'wish[ed] men to be free/As much from *mobs* as kings' *(Don Juan,* IX. 199–200, my italics). Decisions about working-class diet seem to be too important to leave to the working class and – in the work of these two poets born into the most powerful section of the social spectrum – although the voice of 'your labouring people' may be heard, it is unlikely to be given a privileged position.

The part an ascetic, vegetarian diet might play in distinguishing and setting apart a leader in such circumstances is delineated, as Morton points out (p. 83), in the portrait of Conrad in Byron's *The Corsair:*

> Ne'er for his lip the purpling cup they fill,
> That goblet passes him untasted still –
> And for his fare – the rudest of his crew
> Would that, in turn, have passed untasted too;
> Earth's coarsest bread, the garden's homeliest roots,
> And scarce the summer luxury of fruits,
> His short repast in humbleness supply
> With all a hermit's board would scarce deny.
> But while he shuns the grosser joys of sense,
> His mind seems nourished by that abstinence,
> 'Steer to that shore!' – they sail. 'Do this!' 'tis done.
> 'Now form and follow me!' – the spoil is won.
> Thus prompt his accents and his actions still,
> And all obey and few enquire his will.
> *(The Corsair,* I. 67–80)

---

[35] Shelley commented (Note to *Queen Mab, Shelley, Prose,* p. 85): 'It is only the wealthy that can, to any great degree, even now, indulge the unnatural craving for dead flesh, and they pay for the greater licence of the privilege by subjection to supernumerary diseases.'
[36] Burke, *Reflections,* p. 76; *Matthew* 8. 28–33.

Conrad is of Christian rather than Moslem stock – although he is 'more than Moslem' (I. 430) in his abstinence from wine – and his rejection of the food Medora has carefully chosen for him, which includes fruit and sherbet 'in its vase of snow' (I. 427), reinforces his turning away from the orientalized, feminine, softening environment which she and such delicacies symbolize. Similarly, Haidée's pirate father Lambro – ironically described as 'Moderate in all his habits, and content / With temperance in pleasure, as in food' (III. 419–20) – is characterized by the antithesis to Haidée's and Juan's physical indulgence of all sorts, of which the narrator remarks that 'late hours, wine and love are able / To do not much less damage than the table' (III. 527–8). Conrad and Lambro represent a variety of eating (or rather, non-eating) which eschews both British beef-eating bellicosity and oriental softness and feminization, in favour of what Morton terms (p. 83) 'a disciplined and sensitive body: a variety of asceticism used to gain control over others'. This is a style to which Byron himself aspired:

> The simple olives, best allies of wine,
>   Must I pass over in my bill of fare?
> I must, although a favourite 'plat' of mine
>   In Spain, in Lucca, Athens, every where:
> On them and bread 'twas oft my luck to dine,
>   The grass my table-cloth, in open air,
> On Sunium or Hymettus, like Diogenes,
> Of whom half my philosophy the progeny is.
> (*Don Juan* XV. 577–84)

In the last year of Byron's life the association he made between the simple diet and disciplined action and leadership became explicit, and in Ithaca in 1823 he is reported as deliberately choosing to eat unripe grapes, and explaining to the doctor who had advised against them:

> I take them ... in order to accustom myself to any and all things that a man may be compelled to take where I am going – in the same way that I abstain from superfluities, even salt to my eggs, or butter to my bread.
> (Lovell, *His Very Self,* p. 417)

Such asceticism is in sharp contrast with stanza 43 of *Beppo* (1817) where, in a more self-indulgent mood during his first year in Italy, Byron recommends English travellers to stock up on 'Ketchup, Soy, Chilli-vinegar and Harvey' to sauce the bland, undressed fish that the Italians ate during Lent.[37] It echoes the belief, expressed in Plutarch and Rousseau and referred to by Shelley in *A Vindication of Natural Diet,* that the 'extras' served with meat (or fish), such as oil, wine, honey, fish sauce, vinegar and Syrian and Arabian seasonings, added to the detrimental qualities of the flesh: both by disguising that it *was* flesh, which might otherwise repel us, and by breaking down the meat, which made it even harder for the human digestive system to cope with. Such 'superfluities' also reflect the fact that the Britain of Byron's day already had

---

[37] Byron wrote verses in 1811 on 'The Composite merits of Hervey's [sic] Fish Sauce and Hervey's Meditations'.

centuries of foreign trading behind it, and that the British diet of this period comprised several staples which were highly exotic: including one of Byron's favourites, tea (which by then was naturalized among the British upper and middle classes); the raisins which were the essential 'plums' of the plum-puddings Byron enjoyed, and the soy, anchovies and cayenne pepper which were the ingredients of Harvey's and other sauces and which could be obtained only by literally travelling the world from China to French Guyana. The contrasts Byron draws in *The Corsair, Don Juan* and *Beppo* are not only those between different religious cultures, and between fabricated and plain food, but also those between a developed economy reliant on international trade and one based more 'naturally' on local produce.

The Island

Byron's last complete narrative poem, *The Island,* recurs to the belief that purposefulness and masculinity cannot be sustained in environments where the living and loving are too easy. So Torquil's heart, although bred to hardship in the far-northern Hebrides, is tamed by Neuha in far-southern Toobonai

> to that voluptuous state,
> At once Elysian and effeminate,
> Which leaves no laurel o'er the hero's urn.
> (I. 312–14)

Byron's concepts echo those of Montesquieu in *The Spirit of the Laws* (1748), where culture and cultivation are closely related, and the nature of the terrain is perceived to have a defining effect upon the nature of peoples. 'The barrenness of the land makes men industrious, sober, inured to work, courageous, and fit for war', Montesquieu claimed (pp. 278–9):

> they must procure for themselves what the terrain refuses them. The fertility of a country gives, along with ease, softness and a certain love for the preservation of life. ... There are so many savage nations in America because the land itself produces much fruit with which to nourish them.

The Island is Byron's most fully worked-out attempt to present a contemporary civilization totally different from anything he or his readers had known, and in it he constantly uses food to situate the otherness of Toobonai. The poem minimizes the oriental/occidental dimension by making tobacco a uniting feature of Christian and Islamic habits (since 'from east to west' it 'Cheers the tar's labour or the Turkman's rest' [II. 448–9]), while the north/south divide – laboured versus unlaboured eating – is constantly emphasized. On Toobonai, in marked contrast to post-Napoleonic Europe, 'all partake the earth without dispute, / And bread itself is gathered as a fruit'

(I. 215–16).[38] Byron pointed out in a note to the poem that it was specifically to transplant the trees of the 'celebrated bread-fruit' from Tahiti to the West Indies that the voyage on the *Bounty* was undertaken in 1787, leading to the mutiny and the incidents envisioned in his poem; and this tree is imagined as providing the ideologically perfect food:

> The bread-tree, which, without the ploughshare, yields
> The unreaped harvest of unfurrowed fields,
> And bakes its unadulterated loaves
> Without a furnace in unpurchased groves,
> And flings off famine from its fertile breast,
> A priceless market for the gathering guest.
> (II. 260–5)

The fact that the European staple item of diet, bread, was (like meat) hugely labour-intensive to produce, is one of the topics which is frequently featured in the vegetarian discourse of the period. Joseph Ritson, for example, wrote (p. 64), with his eccentric spelling, how:

> The grain of which it [bread] is made, is of all vegetable productions, that which demands most culture, machinery and handleing. Before it is cast into the ground, there must be ploughs to til the ground, harrows to break the clods, dunghils to manure it. When it begins to grow, it must be weeded; when come to maturity, the sickle must be employ'd to cut it down; flails, fanners, bags, barns to trash it out, to winnow it, and to store it up; mils to reduce it to flour, to bolt it, and to sift it; bake-houseës, where it must be kneaded, leaven'd, bake'd and converted into bread. Veryly man never could have existed on the earth if he had been under the necessity of deriveing his first nutriment from the corn-plant.

Shelley, speaking of the change from a meat to a vegetable diet, says:

> The change which would be produced by simpler habits on political economy is sufficiently remarkable. The monopolizing eater of animal flesh would no longer destroy his constitution by devouring an acre at a meal. ... The quantity of nutritious vegetable matter, consumed by fattening the carcass of any ox, would afford ten times the sustenance, undepraving indeed, and incapable of generating disease, if gathered immediately from the bosom of the earth. (Note to *Queen Mab*, Shelley, *Prose*, I. 84)

The Island was written only a few months after Shelley's death, and Charles Robinson regards the poem as Byron's 'final and public tribute to Shelley'.

---

[38] Both the Norman Abbey dinners in *Don Juan* cantos XV and XVI are full of labour: for those who eat and digest them, as well as by implication for those who provide, create and serve them:

> The mind is lost in mighty contemplation
>   Of intellect expended on two courses;
> And indigestion's grand multiplication
>   Requires arithmetic beyond my forces. (XV. 545–8).

It adds to the credibility of Robinson's suggestion that this characteristically Shelleyan vision – of eating which does not antagonistically labour against nature, but relaxedly exists in harmony with it – should be given so prominent a place in the poem.[39] By dint of presenting the story of the poem within a frame which sites it in a time past which no longer exists; by separating out the 'historical' part of the story of the mutineers, and containing the darker, tragic (and, we may think, more realistic) parts of his narrative in and around the figure and death of Fletcher Christian, Byron makes it possible to portray, in Torquil and Neuha, a tale of successful though immature love and to evoke the possibility of an eternally happy ending: 'A night succeeded by such happy days / As only the yet infant world displays' (IV. 419–20).

The bread-tree is second only to the maternal breast, with which it is compared, in providing an instantly-available and constantly-renewed source of food for the childlike people it sustains on Toobonai. Once the mutineers and their pursuers have come and gone – leaving behind only young Torquil to personify the European poet or reader in this Eden – the island is able to return to its pristine state: 'Again their own shore rises on the view, / No more polluted with a hostile hue' (IV. 401–2), and the beneficent natural cycle can begin again. Safe for the while from 'the sordor of civilization' (II. 69) Toobonai is a self-sufficient and (unlike Malthus's) an infinitely sustainable world in which the human need for food is met by a constantly self-renewing nature. Thus it is that, seven years after the debate which gave rise to *Julian and Maddalo*, it is Maddalo who finally creates a full imaginative realization of Julian's 'Utopia': a Shelleyan eco-system which seems capable of lasting for ever.

---

[39] Charles E. Robinson, *Shelley and Byron: The Snake and Eagle Wreathed in Fight* (Baltimore and London: Johns Hopkins University Press, 1976) p. 237: 'I believe we can discover his [Byron's] final and public tribute to Shelley (as well as an exoneration of the 'ocean' that destroyed him) in *The Island; or, Christian and His Comrades.*'

Figure 1: *Portrait of Boatswain*, by Clifton Tomson, 1808, Newstead Abbey Collection NA 245. Byron's much-loved first Newfoundland dog died of rabies in 1808. By courtesy of Nottingham City Museums and Galleries (Newstead Abbey).

Figure 2: *John Wilmot, 2nd Earl of Rochester*, artist and date unknown. Rochester represents the artist in himself through the figure of a monkey, which he crowns with poetic laurels. The ape, like the human artist and the Devil, is seen satirically to mimic the work of God. By courtesy of the National Portrait Gallery, London.

Figure 3: Monument for Boatswain at Newstead Abbey. The extravagant neoclassical monument commissioned by Byron stands in the ruins of the Gothic Abbey at Newstead, and Byron intended to be buried there himself, alongside his dog. By courtesy of Nottingham City Museums and Galleries (Newstead Abbey).

Figure 4: One of twelve illustrations from 'The Wonderful History of Lord Byron and His Dog' written and drawn in 1807 by Elizabeth Pigot, Byron's friend and neighbour at Southwell. It imitates John Harris's *The Comic Adventures of Mother Hubbard and her Dog* (1805). By courtesy of the Harry Ransom Research Center, The University of Texas at Austin.

Figure 5: 'Lyon you are no rogue Lyon'. Frontispiece by Robert Seymour to *The Last Days of Lord Byron* by William Parry, 1825. Byron with his second Newfoundland in Missolonghi, observed by the Firemaster Parry just before Byron's death.

Figure 6: *Dressing the Kitten*, by Joseph Wright of Derby. Educators of the period such as Sarah Trimmer would have disapproved of this scene; believing children should be taught to treat animals correctly: 'neither spoil[ing] them by indulgence nor injur[ing] them by tyranny'. By courtesy of the English Heritage Photographic Library.

Figure 7: *The Four Stages of Cruelty*, engravings by William Hogarth, 1751, Plate 1: 'The First Stage of Cruelty'. Tom Nero (a charity boy of St Giles's Parish, distinguished by the 'S.G.' on his sleeve) tortures a dog while being restrained by a gentleman's son (possibly a portrait of the young Prince George, later George III). In the top left-hand corner, a cat with wings attached is thrown out of a window. © Copyright the British Museum.

Figure 8: *The Four Stages of Cruelty*, engravings by William Hogarth, 1751, Plate 4: 'The Reward of Cruelty'. Tom Nero has been hanged, and his body is being dissected in an operating theatre resembling that of the Royal College of Surgeons. A dog gets its own back on Tom by eating his discarded heart. © Copyright the British Museum.

Figure 9: 'He was overjoyed to find that it sucked as naturally as if it had really found a mother'. Frontispiece by M. Brown illustrating 'The History of Little Jack' by Thomas Day in *The Children's Miscellany*, 1804. The story describes how an abandoned baby, brought up by an old man according to Rousseauian principles, and fed by his nanny-goat, grows up to become a strong and healthy boy. By permission of the British Library (shelfmark CH.790/64).

Figure 10: *The Dog*. Frontispiece by William Blake to William Hayley's *Ballads ... Founded on Anecdotes Relating to Animals*, 1805. Edward is about to swim in a river filled with crocodiles until prevented by Fido the dog. But Fido's brave action costs him his own life, when he falls in himself and is eaten. By permission of the British Library (shelfmark C.58.c.28)

Chapter 5

Animals and Nature: Beasts, Birds and Wordsworth's Ecological Credentials

The critical debate about the nature and extent of Wordsworth's political engagement during (and within the poems of) his 'great period' has been a fierce one. On the one side, New Historicists have claimed that Wordsworth used poetry and transcendence at this time as a means of glossing over his loss of faith in his former radicalism and his move towards the political right. On the other side, critics loosely grouped under an 'ecological' banner have sought to shift the debate away from the question of whether Wordsworth's politics were red or blue by defining the poet's views as proto-ecological, and therefore essentially green.[1]

At the core of this debate is the question of defining Wordsworth's deep involvement with the natural environment, and modern critics have approached this in many different ways. While Marjorie Levinson sees the poet using the landscape as 'a repository for outgrown ego-stages, themselves enshrining certain social values', Alan Bewell argues that 'Wordsworth's turn to nature was not motivated from a desire to avoid or evade politics, but from the belief that nature and the narratives it supports have historically been the very medium of political argument and social control'.[2] Jonathan Bate *(Romantic Ecology,* p. 10) perceives 'not an opposition but a continuity between [Wordsworth's] "love of nature" and his revolutionary politics', while Nicholas Roe emphasizes how 'Well before our own age of environmental crisis, nature spoke to Dorothy, William, and their contemporaries of humankind's social welfare and "moral being"'.[3]

One area of this terrain has, however, been surprisingly little explored. The way in which Wordsworth places and approaches animals in his view of nature seems to have been given very little critical attention, although animals must occupy a central place in any system which seeks to relate humans to the natural environment. It is particularly surprising that no-one from the 'ecological' school of criticism has ventured into this area, since ecology was defined in 1870 by the German zoologist who coined the term, Ernest Haeckel,

---

[1] Jonathan Bate, *Romantic Ecology: Wordsworth and the Environmental Tradition* (London: Routledge, 1991) p. 33: 'The language of *The Prelude* is fleetingly red but ever green.'

[2] Marjorie Levinson, *Wordsworth's Great Period Poems: Four Essays* (Cambridge: Cambridge University Press, 1986) p. 23; Alan Bewell, *Wordsworth and the Enlightenment: Nature, Man and Society in the Experimental Poetry* (New Haven: Yale University Press, 1989) p. 141.

[3] Nicholas Roe, *The Politics of Nature: Wordsworth and some Contemporaries* (Basingstoke: Macmillan, 1992) p. 2.

as a science essentially about animals (and about human beings simply as one species amongst many):

> By ecology, we mean the body of knowledge concerning the economy of nature – the investigation of the total relations of the animal both in its inorganic and its organic environment; including above all, its friendly and inimical relations with those animals and plants with which it comes directly or indirectly into contact – in a word, ecology is the study of all those complex interrelations referred to by Darwin as the conditions of the struggle for existence.[4]

When Wordsworth himself places animals in relation to 'Nature' in Book VIII of *The Prelude,* in a passage almost unchanged between the 1805 and 1850 versions of the poem, he draws the line between man and the natural environment in what seems a surprising place, by locating the beasts and birds *outside* 'Nature'. He learnt to love these creatures, Wordsworth says, not only after the first flush of his youthful passion for Nature was over, but also after he had learnt to love mankind, sometime in his early twenties:

>   a passion, she [Nature],
> A rapture often, and immediate joy
> Ever at hand; he [Man] distant, but a grace
> Occasional, and accidental thought,
> His hour being not yet come. Far less had then
> The inferior creatures, beast or bird, attuned
> My spirit to that gentleness of love,
> Won from me those minute obeisances
> Of tenderness which I may number now
> With my first blessings. Nevertheless, on these
> The light of beauty did not fall in vain,
> Or grandeur circumfuse them to no end.
> (1805 *Prelude*, VIII. 486–97)[5]

At first sight the almost religious intensity of the words Wordsworth uses about animals here – 'light of beauty', 'grandeur', 'circumfuse', 'tenderness', 'obeisances' – might recall the reaching for the sublime in the 'Intimations of Immortality' Ode, but actually the heightened, numinous sound of the passage is not matched by its sense. The poet is in fact firmly *excluding* animals from being a part of the great formative influence on his life, and the high feudal ceremoniousness with which he treats them here is directly opposed to any real power which he allows them to have over him.

The unstraightforwardness with which Wordsworth approaches this area, therefore, and the way in which the critics seem to have by-passed it, have both prompted me to see here the possibility of a repression or denial by the

---

4   Quoted by Bate p. 36, from the translation by Robert P. McIntosh in *The Background of Ecology: Concept and Theory* (Cambridge: Cambridge University Press, 1985) pp. 7–8.

5   In the Dedication to *The White Doe of Rylstone* (line 45), Wordsworth similarly refers to animals as the 'inferior Kinds'. The quotation from Bacon which ends this Dedication elucidates the way Wordsworth uses a 'Chain of Being' model to place animals in a similar relationship to Man as Man stands to God.

imagination of something which is troublesome or difficult to assimilate within the terms which the poet, or the critic, has formulated, and an absence perhaps not unlike that of the beggars or the polluting iron-foundry to which Marjorie Levinson draws attention in 'Tintern Abbey'.[6] One of the propositions explored in this chapter is that animals are problematic – and important – to Wordsworth because they not only lie outside 'Nature', but are also outside 'Man': both inside and outside 'the natural environment', and also both within and without 'culture'. Wordsworth's (and our) conception of them cannot be fully accommodated by any severe distinction between 'the human' and 'the nonhuman', since they are liminal beings which slip and slide according to the context in which they are viewed.[7]

Wordsworth's deployment of the term 'animal' itself indicates some ambivalence. His most frequent use of it is as an adjective (as was characteristic in educated speech up to the eighteenth century) and he applies it in this way to describe the non-spiritual part of *human* consciousness – the 'glad animal movements' of childhood, for example, or the 'animal delight though dim' felt by an old man who has suffered a stroke.[8] When he deploys 'animal' as a substantive noun – as was more and more common from the seventeenth century onwards – he often qualifies it and its cognates with an adjective which separates it from Man and renders it silent and inarticulate: such as are the 'dumb animals' saved and cared for by the shepherd Michael (line 71); the 'mute Animal' which gives its name to *The White Doe of Rylstone* (line 876); the 'dumb partner' of 'Hart-Leap Well' (line 38), and the 'dumb creatures' of *The Prelude* (1805 *Prelude,* V. 303).

---

6   See Levinson, pp. 1–57.

7   See Grevel Lindop, 'The Language of Nature and the Language of Poetry' in *Wordsworth Circle* XX. 1 (Winter 1989) pp. 2–9 (p. 3): 'For us, the boundaries of nature shift depending upon where we choose to draw our circle round the human. If we decide that the truly human is the mind, then the body and all that surrounds it becomes nature. If we decide that the human represents human beings with their realm of cultural artefacts, then nature becomes whatever seems to be beyond that realm.'

8   'Tintern Abbey' 74; 'The Matron of Jedborough and her Husband' 62. The 1797 edition of the *Encyclopaedia Britannica,* which was used and later owned by the Wordsworths, offers a useful encapsulation of the deployment of the word in this period: 'ANIMAL in natural history, an organized and living body, which is also endowed with sensation: thus, minerals are said to grow or increase, plants to grow and live, but animals alone to have sensation. ... Animal, used adjectively, denotes any thing belonging to, or partaking of, the nature of animals. Thus, animal affections are those that are peculiar to animals; such are sensation and muscular motion' (II. 21–4). Moving from these essentially Aristotelian definitions, the *Encyclopaedia* rather bizarrely suggests that the 'principle of self-preservation' is also 'a true characteristic of animal life': 'The power of vegetation, for instance, is as perfect in an onion or a leek, as in a dog, an elephant, or a man; and yet, though you threaten a leek or an onion ever so much, it pays no regard to your words, as a dog would do; nor, though you would strike it, does it avoid a second stroke' (II. 22). *The Encyclopaedia Britannica; or, A Dictionary of Arts, Sciences and Miscellaneous Literature,* 3rd edn, 18 vols (Edinburgh: A. Bell and C. Macfarquhar, 1797). For the Wordsworths' ownership and use of this encyclopaedia, see Duncan Wu, *Wordsworth's Reading 1800–1815* (Cambridge: Cambridge University Press, 1995) p. 85, and Chester L. Shaver and Alice C. Shaver, *Wordsworth's Library: A Catalogue* (New York: Garland, 1979) p. 88.

This silence is a conundrum for a poet deeply concerned with the nature of the poetic voice and with speaking and articulation in all its forms. Because animals have consciousness and volition and yet do not speak, they occupy the margin between 'the mute' and 'the brute': between the electively and the non-electively silent. It appears, in fact, that the significance of animals lies for Wordsworth in an area which combines two of the most powerful preoccupations of his work: the interaction between humanity and nature, and the development of his own ability to articulate the meaning of the world as a poet. Beasts and birds fascinate him because, unlike the inorganic and inanimate objects upon whom the poet may bestow a voice of his own choosing, they are creatures which can (non-verbally but sometimes vocally) speak for themselves. They challenge him because they remind him of the fragility of his own ability to articulate meaning and because they bring into play his conviction that fully 'natural' language – the language which Jacob Boehme and David Hartley identified with an Adamic, pre-Babel tongue – is beyond or above words anyway, and is to be found in the 'mute, insensate things' of the Lucy poems.[9] There is a constant sense of struggle between Wordsworth's sensitivity to the voiceless objects he contemplates, and his egotistical wish to speak for them: exhibiting what Will Christie calls both 'the yearning after a meta-language that will resist analysis, yet be perfectly expressive' and also 'a persistent fear that his audience would misread the language of nature,' which led to his 'dramatising and asserting his own deeply felt reading.'[10]

Marginals and inanimates

This is a dichotomy also to be found in the poet's encounters with the 'marginal' human individuals, such as idiots, children, villagers, old women, vagrants, the blind, the deaf and the mute, who form the subjects of many of his 'great period' poems. Jonathan Wordsworth, for instance, opens his study of Wordsworth's 'borderers' with a demonstration of how the poet sees animals, as well as marginal human beings, as inhabitants of a 'hinter-world' between different states. He groups together the leech-gatherer of 'Resolution and Independence' – who shifts from human being to geological feature, thence to animal, and then back to human being again – and the sleeping horse of a rejected *Prelude* passage, who is

> Like an amphibious work of Nature's hand,
> A borderer dwelling betwixt life and death,
>   A living statue or a statued life.[11]

---

[9] See George Steiner, *After Babel: Aspects of Language and Translation,* 2nd ed (Oxford: Oxford University Press, 1992) p. 65, and Lindop, p. 3. 'Three Years She Grew' (line 18).

[10] Will Christie, 'Wordsworth and the Language of Nature' in *Wordsworth Circle,* XIV. 1 (Winter 1983) pp. 40–47 (p. 43).

[11] Jonathan Wordsworth, *William Wordsworth: The Borders of Vision* (Oxford: Clarendon Press, 1982) pp. 1–3.

Similarly, in his study of the poet's use of Enlightenment scientific discourse, Alan Bewell provides a description of the Wordsworthian Idiot which could equally well be applied to the poet's conception of animals: to, for instance, the 'mild and good pony' in 'The Idiot Boy' as much as to Johnny himself. The Idiot, Bewell says (p. 57), is one who occupies 'the threshold between nature and man, and could be seen as a figure linking the two states'. Bewell shows (p. 40) how Wordsworth's revisionary reworking of the primitive encounter which is a feature of eighteenth-century anthropological discourse draws attention to and privileges the silence of these marginal figures: the blind beggar and Johnny, he says, 'promise to clear up human mysteries that have been hidden since the beginning of time. But such expectations are never fulfilled'.[12] As with marginal human beings, therefore, Wordsworth's accounts of encounters with animals often seem frustrated, awkward and oblique, and this is an area in which the reader becomes more than usually aware of the various kinds of literary artifice which have been deployed. The poet's wish to make the animal 'speak' is almost always 'processed' in some way, so that the 'speaking' is not directly recounted, but mediated through an interpretative mechanism.

This is in marked contrast to the ease with which Wordsworth enables totally *in*animate objects to have a voice. In the same way that Coleridge's icicles, in 'Frost at Midnight', can 'shine' with what is actually *reflected* moonlight, Wordsworth's hills, rocks, rivers and other physical objects are made to express directly and relatively unproblematically the emotion which the poet invests in them (and this is so whether the feeling is perceived to have been received from them or projected onto them in the first place). Among the first influences on the poet recounted in *The Prelude* is the River Derwent which 'sent a voice/That flowed along my dreams' and made 'ceaseless music' which 'composed my thoughts / To more than infant softness' (1805 *Prelude*, II. 275–82). When birds-nesting as a child on the high crags, he notes the 'strange utterance' with which 'the loud dry wind' blows through his ears (1805 *Prelude*, I. 348–9); and in his memories of the 'spot of time' just before his father's death, when as a boy he perched on a crag beside a single sheep, it is the sounds made by the inanimate – rather than the animate – components of that moment that come vividly back to him many years later: 'the bleak music of that old stone wall,/The noise of wood and water' (XI. 378–9).

In Book II of *The Prelude* the effect of the voice of inanimate nature upon the poet becomes explicit as he recounts how as a youth he

> felt whate'er there is of power in sound
> To breathe an elevated mood, by form

---

[12] Bewell (p. 40) interprets the silence as representing 'Wordsworth's resistance to the textual marginalization that these individuals underwent during the previous century – in short, his attempt to write the Enlightenment discourse on marginality out of existence by seeking to undo its pleasure in producing marginals and taking rhetorical advantage of those who cannot speak or suffer from physical or mental hardships.' Wordsworth wrote to John Wilson in 1802, in response to the latter's criticism of 'The Idiot Boy': 'I have often applied to Idiots, in my own mind, that sublime expression of scripture that, "their life is hidden with God."' *The Letters of William and Dorothy Wordsworth*, ed by Ernest de Selincourt, 2nd edn, vol. I, *The Early Years 1787–1805*, rev. by Chester L. Shaver (Oxford: Clarendon Press, 1967) p. 356.

> Or image unprofaned; and I would stand
> Beneath some rock, listening to sounds that are
> The ghostly language of the ancient earth,
> Or make their dim abode in distant winds.
> (1805 *Prelude*, II. 324–9)

Later, he wishes for 'a music and a voice' as harmonious as the 'breezes and soft airs', the 'brooks muttering along the stones' and the 'groves', 'that I might tell / What ye have done for me' (1805 *Prelude*, XI. 10–22), and when in Book III he describes himself at Cambridge he shows how habitual it had become for him to give a life to inanimate natural objects:

> To every natural form, rock, fruit, or flower,
> Even the loose stones that covered the highway,
> I gave a moral life – I saw them feel
> Or linked them to some feeling. The great mass
> Lay bedded in a quickening soul, and all
> That I beheld respired with inward meaning.
> (1805 *Prelude* III. 124–9)

The inanimate objects of a personified Nature can be relied upon never to betray the heart that loves them. The animate ones, by contrast, retain a stubborn ability to bamboozle the poet by withholding their voices: as they do from the boy of Winander who blows through his hands in 'mimic hootings to the silent owls',

> And they would shout
> Across the wat'ry vale, and shout again,
> Responsive to his call, with quivering peals
> And long halloos, and screams, and echoes loud
> Redoubled and redoubled – concourse wild
> Of mirth and jocund din! And when it chanced
> That pauses of deep silence mocked his skill,
> Then sometimes in that silence, while he hung
> Listening, a gentle shock of mild surprize
> Has carried far into his heart the voice
> Of mountain torrents . . .
> (1805 *Prelude*, V. 399–410)

Objects which have been given a voice by the poet prove generally to be more reliable and, as Jonathan Bate has remarked, it is often in the moments when the animals *cease* to cry or speak that the poet seems to find most meaning, or when their expression has undergone a metamorphosis from a living creature's cry to the voice of an inanimate object.[13] This is so with the lamb in Book IV of *The Excursion,* whose voice is 'sent forth / As if the visible mountain made the cry,' so that:

---

[13] 'Wordsworth's world at its moments of intensity is strangely silent: the vision enters the boy when the owls don't reply': Jonathan Bate, *The Song of the Earth*, p. 186.

              from out the mountain's heart
The solemn voice appeared to issue, startling
The blank air – for the region all around
Stood empty of all life, and silent
Save for that single cry, the unanswer'd bleat
Of a poor lamb – left somewhere to itself,
The plaintive spirit of the solitude!
(*The Excursion,* IV. 402–12)[14]

Animals do feature as a part of Wordsworth's reverential communion with nature:

I felt the sentiment of being spread
O'er all that moves, and all that seemeth still, ...
O'er all that leaps, and runs, and shouts and sings,
Or beats the gladsome air, o'er all that glides
Beneath the wave, yea, in the wave itself
And mighty depth of waters. Wonder not
If such my transports were, for in all things
I saw one life, and felt that it was joy.
(1805 *Prelude,* II. 420–30)

Despite his 'transports' about the 'one life', however, the poet finds his own participation in it easier once the *actual* sight and sound of the animals has been removed and he can recall them to his 'inward eye' and inward ear:

One song they sang, and it was audible –
Most audible then when the fleshly ear,
O'ercome by grosser prelude of that strain,
Forgot its functions and slept undisturbed.[15]
(II. 431–4)

Enjoyment

The belief that what he is perceiving in the movements or cries of animals is a reflection of their enjoyment of life is frequently expressed in Wordsworth's work, and this apparent joy can be variously deployed as a means of intensifying the poet's own happiness or to provide a contrast with human

---

[14] The Wanderer is interested in and concerned for animals, and Book IV of *The Excursion* also contains a long description of his observation of the various 'creatures that in communities exist / Less, as might seem, for general guardianship ... Than by participation of delight / And a strict love of fellowship, combined' (IV. 440–4).

[15] Douglas B. Wilson, in *The Romantic Dream: Wordsworth and the Poetic of the Unconscious* (Lincoln: University of Nebraska Press, 1993) pp. 18–19 discusses Coleridge's and Wordsworth's ideas about 'the tyranny of the senses' and throws light on Wordsworth's use of the word 'slept' here. While awake, a person falls prey to many sensory distractions, but during sleep this process is reversed, and nocturnal images benefit from this sensory deprivation. In Coleridge's words: 'From the exclusion of all outward impressions on our sense the images in sleep become proportionally more vivid than they can be when the organs of sense are in their active state.'

misery. In *Home at Grasmere,* it is Wordsworth's own pleasure in spiritual homecoming which seems to be responsible for the pathetic fallacy by which even the raven's croak becomes 'a strange sound of genial harmony' (MS D, 582), and by which the wheeling of flocks of birds over the lake is characterized as a 'jubilant activity' (212).[16] In 'Lines written in Early Spring' (1798), and in 'Resolution and Independence' and the 'Intimations of Immortality' Ode (both 1802), the 'happy children of earth', 'blissful creatures' and the beasts who 'keep holiday' anticipate the arguments of Natural Theology, as expressed by William Paley in 1802:

> It is a happy world after all. The air, the earth, the water, teem with delighted existence. In a spring moon, or a summer evening, on whichever side I turn my eyes, myriads of happy beings crowd on my view. 'The insect youth are on the wing.' Swarms of new-born *flies* are trying their pinions in the air. Their sportive motions, their wanton mazes, the gratuitous activity, their continual change of place without use or purpose, testify their joy, and the exultation they feel in their lately discovered faculties. ... Other species are *running about,* with an alacrity in their motions, which carries with it every mark of pleasure. ... If we look to what the *waters* produce, shoals of fry of fish frequent the margins of rivers, of lakes, and of the sea itself. These are so happy that they know not what to do with themselves. Their attitudes, their vivacity, their leaps out of the water, their frolics in it (which I have noticed a thousand times with equal attention and amusement), all conduce to show the excess of spirits, and are simply the effects of that excess. ... what a sum, collectively, of gratification and pleasure have we here before our view!*(Natural Theology,* pp. 490–2)

But although Wordsworth's 'Lines' similarly celebrate non-human joy:

> The birds around me hopped and played,
> Their thoughts I cannot measure; –
> But the least motion which they made,
> It seemed a thrill of pleasure,
> ('Lines written in Early Spring', 13–16)

they also lament the way in which humanity seems to be at odds with it, and therefore question Paley's basic premise:

> If this belief from heaven be sent,
> If such be Nature's holy plan,
> Have I not reason to lament
> What man has made of man?
> (21–4)

In *Peter Bell* Wordsworth appears deliberately to confront his own difficulties with articulate animals, by giving the communication between man and animal a central place in the story and making dramatic use of an animal's cry: not a joyful one this time, but 'The hard dry see-saw of [the Ass's]

---

[16] *Home at Grasmere: Part First, Book First, of* The Recluse *by William Wordsworth,* ed by Beth Darlington (Ithaca, New York: Cornell University Press, 1977) pp. 89 and 55.

horrible bray' (480). This may reflect the fact that the poem, written in 1798, seems to have been intended as a companion piece to Coleridge's 'Ancient Mariner', and shares that poem's interest in the humanizing power of beasts and birds.[17] *Peter Bell* was not published until 1819, but Wordsworth's concerns in it reflect the politics of the turn of the century when there was a close association between the interests of humanitarianism and those of animal welfare. As I have shown in Chapter Three, the efforts of William Wilberforce and his followers to abolish the slave trade and improve the lot of agricultural workers were mirrored by their attempts to introduce Parliamentary measures to ban bull-baiting in 1800 and 1802. Keith Thomas comments (pp. 184–5) on how

> In practice, it was almost impossible to reflect on animals without being distracted by the conflicting perceptions imposed by social class. ... In the seventeenth and eighteenth centuries much of the pressure to eliminate cruel sports stemmed from the desire to discipline the new working class into higher standards of public order and more industrious habits. It is often remarked (and it was noticed at the time) that it was the sports with a strong proletarian following which were outlawed.

By the time *Peter Bell* was published, however, more than two decades later in the year of Peterloo, these connections seemed outdated and naive, and the poet appeared to be using the subject matter of childish and unsophisticated fables to try to deal with a serious political problem: subjecting the dangerous and dispossessed underclass of vagrants, epitomized by Peter Bell, to what Leigh Hunt called 'reformation by harebells and hell-fire', and leading Shelley, in *Peter Bell the Third,* to deride the poem as 'the folly which soothes tyranny'.[18]

---

[17] The two poems illustrate well the division of labour Wordsworth and Coleridge planned for *Lyrical Ballads:* 'Coleridge was to write about "persons and characters supernatural, or at least romantic," but though supernatural they had to possess "a semblance of truth sufficient to procure ... that willing suspension of disbelief for the moment, which constitutes poetic faith". Wordsworth, for his part, was "to give the charm of novelty to the things of every day, and to excite a feeling analogous to the supernatural, by awakening the mind's attention to the lethargy of custom, and directing it to the loveliness and the wonders of the world before us."' R. L. Brett and A. R. Jones, *'Lyrical Ballads': Wordsworth and Coleridge. The Text of the 1798 edition with the additional 1800 poems and the Prefaces edited with introduction, notes and appendices* (London and New York: Routledge, 1963 (2nd edn 1991, repr. 1993) p. xxii. The quotations are from *Biographia Literaria*, ed by J. Engell and W. J. Bate, 2 vols, *The Collected Works of Samuel Taylor Coleridge,* Bollingen Series, (London: Routledge and Kegan Paul, 1983) II. 6–7.

[18] *The Examiner* number 592 (2 May 1819) p. 282. The curse of dreadful Dullness which falls upon Peter in Shelley's poem appropriately makes the animals suffer too:

> The birds and beasts within the wood,
> The insects and each creeping thing,
> Were now a silent multitude;
> Love's work was left unwrought – no brood
> Near Peter's house took wing. ...
> No jackass brayed; no little cur
> Cocked up his ears. (748–56).

The Ass's gestures with its eyes and ears and body, its grin and its bray were among the features which exposed Wordsworth to mockery and to parody, especially in passages such as that in which the animal tries to draw Peter's attention to the body of its dead master in the water by using a kind of sign-language:

> 'Twas but one mild, reproachful look,
> A look more tender than severe;
> And straight in sorrow, not in dread,
> He turned his eye-ball in his head
> Towards the smooth river deep and clear.
> (436–40)

These cultural gestures are inherently unconvincing when supposed to be part of a real animal's vocabulary, however thoroughly domesticated it might be.[19] Leigh Hunt suggested that the Learned Pig – who is among the exploited performing animals mentioned in *The Prelude*'s description of London (1805 *Prelude,* VII. 681) – should be brought in to interpret the Ass's 'horrible grin'. Wordsworth makes Peter an animal-like human, through references to his 'unshaped half-human thoughts' (296), his 'dark and sidelong walk', his 'long and slouching' gait (306–7) and 'his forehead wrinkled ... and furred' (311). But the poet's attempt, conversely, to make one of the agents of Peter's humanization an animal with human characteristics risks tipping the story into a less dignified genre – that of the fable or children's story, where the beasts can think and behave and even speak in human ways – and this weakens the coherence of the work.

In another poem of this period – 'The Pet-Lamb, a pastoral', written for the 1800 edition of the *Lyrical Ballads* – Wordsworth does contrive to give expression to an animal without losing its (or the poet's) dignity, by making a child the interpreter between the animal and the narrator (as the latter is also a self-conscious interpreter between the reader and the child). The narrator is much less tentative about appropriating the child's, actually unspoken, expression ('but half of it was hers and one half of it was *mine*' [64] Wordsworth's italics), than the child is about interpreting the lamb's inarticulate behaviour:

> It will not, will not rest! – poor Creature can it be
> That 'tis thy Mother's heart which is working so in thee?
> Things that I know not of belike to thee are dear,
> And dreams of things which thou can'st neither see nor hear.
> (49–52)

Altogether, however, the poem conveys well by its medium what is also its message. Like 'Anecdote for Fathers', one of its subjects is the refraining from forceful obtrusion and the delicacy of observation that works best when

---

[19] See Jeremy Bentham's comment, however (mentioned in Chapter One) that 'a full-grown horse or dog is beyond comparison a more rational, as well as a more conversable animal than any infant of a day, or a week, or even a month, old.'

sensing and translating consciousness between adults and children, and between human beings and other species.

Dorothy Wordsworth

As Susan M. Levin points out, Dorothy Wordsworth's prose story 'Mary Jones and her Pet-Lamb' 'most explicitly intersects with William's poem', and 'the story and the poem complement each other to provide an interesting demonstration of some of the differences between Dorothy's imagination and that of her brother'.[20] A comparison with Dorothy Wordsworth's approaches to animals in both her prose and her poems does in general point up the oddness, unconventionality and originality of the way William sometimes writes about birds and beasts; and this is an area where, in contrast to the way in which the siblings' work is usually characterized, it is Dorothy who achieves a more outward-reaching and generalizing vision, and William's verse which may appear idiosyncratic and preoccupied with the working out of personal concerns and relationships. While William, seeking to appropriate nature to create a myth of self, often seems to be struggling to know where to place animals within this myth and within 'Nature', Dorothy more simply sees both animals and herself as an undisputed part of Nature, for good or ill.

Thus 'The Floating Island at Hawkshead', before it sinks, provides a microcosm which encompasses all aspects of the natural world: where 'warbling birds their pastime take', where 'berries ripen, flowerets bloom', and where 'insects live their lives – and die' (12–15). The imaging of the island as 'A peopled world ... in size a tiny room' (16) connects the natural world with the indoor living-space to which Dorothy was confined, and demonstrates how both may be threatened and destroyed, as well as nurtured, by Nature.

Dorothy's 'Loving & Liking: Irregular Verses Addressed to a Child' shares her brother's preoccupation with taxonomizing the natural world into different categories of being, demonstrated both in the *Prelude* passage quoted earlier in this chapter, and in the *Essay on Epitaphs* cited in Chapter One. But whereas William makes careful distinctions between the degrees of reverence to be accorded respectively to 'Man', 'Nature' and 'The inferior creatures, beast or bird', Dorothy brings together all living things – both animal and vegetable, and ranging from members of the child's family to the strawberry growing on its plant – into the single category of those one may 'love':

> Nor blush if o'er that heart be stealing
> A love for things that have no feeling
> Nor can repay, by loving you,
> Aught that your care for them may do.
> (27–30)

In their dead, plucked or cooked form, in which they may be eaten, fruits and animals both become merely 'likeable'. Live frogs and toads are introduced as creatures which *should* be loved, and not shuddered over, (a popular motif at

---

[20] Susan M. Levin, *Dorothy Wordsworth & Romanticism* (London: Rutgers State University, 1987) p. 53. Dorothy Wordsworth's poems are also quoted from this source.

this time in children's books), and the frog who 'swims by perfect law of Nature', and is observed 'Glancing amid the water bright/ And sending upward a sparkling light', is 'A model for a human creature' (17–20).

All living things are thus part of a single continuum for Dorothy, encompassing flowers, trees, birds and her own being, and when she calls up the natural world in 'Thoughts on my sick-bed', she remembers them all as equally part of nature:

> The violet betrayed by its noiseless breath,
> The daffodil dancing in the breeze,
> The carolling thrush, on his naked perch,
> Towering above the budding trees.
> (17–20)

Like Coleridge, Dorothy often identifies particularly with birds, although as Levin points out (p. 33), her identification with natural objects in general is strongly metonymical, and therefore her work is not specific about the connections between self and bird, as Coleridge is in his more self-assertive metaphorical or symbolic mode. The best-known of Dorothy's identifications of this type is with the nesting swallows, whose attempts and failures to build a nest on their cottage she painfully chronicles in her Grasmere Journal while William is away courting his future wife. Her relationship with the birds, psychologically and physically, could scarcely be closer, as she watches them for hours at her bedroom window 'with their soft white bellies close to the glass, and their forked, fish-like tails'.[21] When their nest falls, it is not difficult to read this as a symbol for Dorothy of the way in which hers and William's 'sweet' life together must be destroyed by his impending marriage. Soon the nest is being reconstructed, but at this point Dorothy is obliged to leave home to visit her future in-laws, before she can see whether this time it will become a home in which the birds can successfully raise a family. Once again, though this time in prose, the grammatical construction of her evocation of the world she must leave behind barely differentiates between animate and inanimate objects in their ability to 'sing':

> The Swallows I must leave them the well the garden the Roses, all.
> Dear creatures!! They sang last night after I was in bed – seemed to be
> singing to one another, just before they settled to rest for the night.
> Well, I must go. Farewell —— (p. 146).

William Wordsworth's version of 'The Pet-lamb' explores, as I have discussed, the difference between a poet's perception and that of a child: the composing of the poem addresses the poet's rather esoteric need to define different categories of creative imagination. Dorothy's prose story on the same theme, by contrast, aims to meet the emotional needs of the child to whom it is told: it first evokes a secure familial environment which is threatened by the loss of the lamb and the child, and then provides most satisfyingly for its re-establishment when child and lamb are found safe, asleep together on the

---

[21] *Journals of Dorothy Wordsworth* ed by Mary Moorman (London: Oxford University Press, 1971) p. 137.

mountain. The text of the tale is highly referential, alluding overtly to the famous story of the Babes in the Wood, cared for (or, in some versions, buried) by robins who strow leaves on them, and invoking the ancient notion, emphasized by authors from Pythagoras to Aquinas, and from Locke to Hogarth, that children learn how to relate correctly to society by caring for animals:

> Mary Jones (having no Brothers or Sisters) had not any Companions like herself; yet she had playfellows; the great Sheep-dog fondled, & jumped about her; & she fed the Chickens & the Ducks, & they would flock round her whenever she went into the yard; and nobody had such sleek beautiful cats as little Mary. ... she had pigeons, too, & they would peck out of her hand, and she loved to sit and watch the bees carry their burthens to the hive; she was not afraid of them, for, as she did not tieze them, they never stung her. ... [Mary] was a very dutiful and good Girl, and continued to be a comfort to her Parents, and after she was grown up to be a Woman, took care of them when they were infirm and old.. ('Mary Jones and her Pet-lamb', from Levin, pp. 239 and 241)

There is also a more up-to-date, Rousseauian, dimension to Dorothy's work for children, alongside the more traditional, Lockean, one. Dorothy and William partly followed Rousseauian methods in their upbringing of little Basil Montague, and Dorothy explained *(Wordsworth Letters,* 1787–1805, p. 180) to Jane Marshall in March 1797: 'Our system respecting Basil ... is a very simple one, so simple that in this age of systems you will be hardly likely to follow it. We teach him nothing at present but what he learns from the evidence of his senses. ... He knows his letters, but we have not attempted any further step in the path of *book learning.* Our grand theory has been to make him happy in which we have not been altogether disappointed.' This 'system' is reflected in her verse, so that the younger child in 'The Mother's Return' is seen to possess an enviable 'oneness' with animal nature:

> *She* wars not with the mystery
> Of time and distance, night and day,
> The bonds of our humanity.
>
> *Her* joy is like an instinct, joy
> Of Kitten, bird, or summer's fly;
> She dances, runs, without an aim
> She chatters in her ecstasy.
> (17–23)

Dorothy's writing for children demonstrates her knowledge of the way contemporary educational theory and practice deployed animals as subject-matter and object-lessons for children. She also indicates her consciousness of the wider political debate about animals and rights: reflecting, for example, on the way in which the use of domestic animals might be seen as exploitative or even cruel:

> The steed, now vagrant on the hill,
> Rejoices in this sacred day,

Forgetful of the plough – the goad –
And, though subdued, is happy as the gay.

Conscious that they are safe from man
On this glad day of punctual rest,
By God himself – his work being done –
Pronounced the holiest and the best.
('Lines written ... on the morning of Sunday April 6th', 13–24)

In general, Dorothy seems as well-informed as William about the
contemporary political, religious and other arguments about the status of
animals; and in the personal dimension of their respective works – that where
deep-seated, individual responses are concerned – Dorothy's work provides
evidence of an attitude to animals that is more accepting, more reposeful, less
striving and less anxious – but also far less ambitious in scale – than is that of
her brother.

The doe and the hart

There are two other poems of William's so-called 'great period' or 'great
decade' in which he uses an animal as a major protagonist in the story. 'Hart-
Leap Well' opened the volume of new poems in the 1800 edition of *Lyrical
Ballads,* and *The White Doe of Rylstone; or, The Fate of the Nortons* (written
between 1807 and 1808) was published on its own in an expensive quarto
edition of 1815. Both poems were, then, ones which Wordsworth set
considerable store by, and he later described *The White Doe* as 'in conception,
the highest work he had ever produced.'[22] In neither of these poems does
Wordsworth repeat the attempt to humanize the animal as in *Peter Bell,* but the
Hart and the Doe are nevertheless made highly expressive. Their
expressiveness, moreover, is of a kind which approaches a spiritual
symbolism, which matches in this respect the symbolic use of animals by
Spenser and Marvell, to which Wordsworth seems to be alluding.[23]

---

[22] Christopher Wordsworth, *Memoirs of William Wordsworth.* 2 vols (London: Ticknor,
Reed and Fields, 1851) 2. 313.

[23] See *The Faerie Queene,* I. i. 28–36

A lovely Ladie rode him faire beside
Upon a lowly Asse more white than snow,
Yet so much whiter, that the same did hide
Under a vele, that wimpled was full low,
And over all a blacke stole she did throw,
As one that inly mournd: so was she sad,
And heavie sat upon her palfrey slow;
Seemed in heart some hidden care she had,
And by her in a line a milke white lambe she lad.

The image is interpreted in *The Works of Edmund Spenser: A Variorum Edition,* ed by Edwin
Greenlaw, Charles Grosvenor Osgood and Frederick Morgan Padelford (Baltimore: Johns
Hopkins Press, 1932), p. 178, as meaning 'Innocence led by Truth', with Una representing
Truth or the Reformed Church, and with a reference to *Isaiah* 49. 23. By contrast,
Wordsworth's symbolism in *The White Doe* refers to the Roman Catholic Church. Wordsworth

There are similarities between Wordsworth's two works in the way the narrative is heavily framed, recalling how the animal's consciousness is approached by a series of stages in 'The Pet-Lamb'. In 'Hart-Leap Well' the frames are provided by an introductory prose note, by the ballad-like story of Sir Walter's hunt and then, in the second part, by the narrative of the poet and the shepherd's story. The only sound actually uttered by the Hart itself in the poem is 'the last deep groan his breath had fetched' (43), but the creature's natural inarticulateness – both because it *is* an animal and because it is dying – is used as an opportunity to construct human and non-human expressions of meaning of contrasting sorts about its death. These include the cheerful, insensitive self-congratulations of the hunting knight and the hesitant, self-effacing story of the shepherd, which itself incorporates the speeches of several others: including the 'some' who speculate about the current state of the well, the 'arbour' which 'does its own condition tell' (129), the water itself which 'send[s] forth a dolorous groan' (136) and the shepherd's own hypothetical reconstruction of the Hart's dying consciousness:

What thoughts must through the creature's brain have pass'd! ...
For thirteen hours he ran a desperate race;
And in my simple mind we cannot tell
What cause the Hart might have to love this place,
And come and make his death-bed near the well.
(141–7)

The narrator's concluding message, which is strongly consonant with modern ecological thinking, reiterates in abstract, educated speech the message that has already been stated (we may think, more effectively) by the shepherd's simple speech, by the animal's dying groan and by inanimate nature in its unwillingness to visit the place:

One lesson, Shepherd, let us two divide,
Taught both by what she [Nature] shews, and what conceals,
Never to blend our pleasure or our pride
With sorrow of the meanest thing that feels.
(177–80)[24]

---

wrote to Walter Scott on 7 November 1805 that he had not seen Andrew Marvell's poems 'these many years'. But he may have looked at them again between 1805 and 1807, since the echoes of 'The Nymph complaining for the death of her Faun' in *The White Doe* seem too strong to be fortuitous. See *Andrew Marvell*, ed Frank Kermode and Keith Walker (Oxford: Oxford University Press, 1994).

[24] Shelley drew ironic attention to this passage in *Peter Bell the Third* (VI. 584–8), contrasting it in a footnote with the 'description of the beautiful colours produced during the agonizing death of a number of trout' in *The Excursion* (VIII. 556–71). 'That poem,' Shelley said, 'contains curious evidence of the gradual hardening of a strong but circumscribed sensibility, of the perversion of a penetrating but panic-stricken understanding'. Since Shelley quotes the 'Hart-Leap Well' passage, he seems to be referring specifically to Wordsworth's 'sensibility' to animals and the natural environment. Byron reworked the image of the changing colours of the dying fish in *Childe Harold* IV. 258–61:

parting day
Dies like the dolphin, whom each pang imbues

'Hart-Leap Well' makes relatively simple use of this device of inventing and deploying various voices, but the same essential configuration – of a narrating persona who may or may not be the poet, faced with another voice or voices which need to be more or less deliberately interpreted – is characteristic of many of Wordsworth's poems of this period. The technique requires the poet/narrator to use a language which, as John Barrell defines it, 'can only be spoken by a fully autonomous and individual subject, a subject who is male, educated, fully differentiated from sense and nature'.[25] This voice may itself be treated ironically – 'I've measured it from side to side / 'Tis three feet long and two feet wide' (The Thorn, stanza 3, early variant) – while the other voices in the poem may speak a variety of languages appropriate to those who are imperfectly differentiated from nature. Barrell draws attention (p. 143) to the way in which Dorothy Wordsworth, for instance, in 'Tintern Abbey' may or may not be judged 'capable of uttering or understanding the language of meditation in the poem'. This differentiation draws on the distinction I have discussed in Chapter One in the liberal, republican ideology of the eighteenth century between property-holding, politically-participating, male citizens, with both the ability to bear arms and the capacity to own land and thus to have a physical stake in the welfare of the state, and the groups of subjects who mediated between the citizens and nature, providing their material needs. These latter groups – who shared to a greater or lesser degree in a feminized representation of nature and who were therefore characterized ambivalently (as innocently unsophisticated yet dangerously volatile) – were women, the working classes and colonized races. To this continuum, in terms of Wordsworth's poetry, one might add animals: which, as we have seen, he characterized as outside and distinct from 'Nature'.

The White Doe of Rylstone shares 'Hart-Leap Well's elaboration of framing devices, including the 'In trellised shed' dedication; and, like 'The Thorn', it opens with a babel of voices, each 'interpreting' the appearance of the Doe at Bolton Abbey in a different way. These are soon cleared away in order to make room for the narration of the poet, who apparently alone can explain the silent creature's 'true' meaning. However, the poet's delicacy is such that – although he will not allow the simpler interpretations to stand – he seems unwilling to mark the white immanence of the Doe, or the virginality of the heroine Emily, with the crudeness of an overt explanation, and the loading of the animal and the woman with significance is matched by a rapt silence about what that significance actually is. This struck even the poet's friends as unwarranted religiosity or mysticism – if not outright mystification – and Charles Lamb reacted irreverently to the poem in manuscript with a whole butcher's block of fleshly metaphors:

With a new colour as it gasps away,
The last still loveliest, till – 'tis gone – and all is grey.
[25] John Barrell, 'The Uses of Dorothy: "The Language of the Sense" in "Tintern Abbey",' in New Casebooks: Wordsworth, ed by John Williams (Basingstoke: Macmillan, 1993) p. 150.

No alderman ever longed after a haunch of buck venison more than I
for a Spiritual taste of that white Doe you promise. I am sure it is
superlative, or will be when *drest,* i.e. printed –. All things read *raw* to
me in MS.[26]

Not surprisingly, Wordsworth became somewhat defensive in recounting to
Coleridge his explanation of how the symbolism of the poem was to be read:

I said to him [Lamb] further that it could not be popular because some
of the objects and agents, such as the Banner and the Doe, produced
their influences and effects not by powers naturally inherent in them,
but such as they were indued with by the Imagination of the human
minds on which they operated.[27]

Wordsworth's assignment here of the cause of meaning in the Doe to 'the
Imagination of human minds' indicates his own consciousness of one of the
mechanisms he deployed widely himself when allocating the significance or
expression of animals in poetry. It was also a convenient answer to Walter
Scott, who had pointed out in 1808 that the story the poem tells is, in terms of
historical authentication, entirely spurious.[28]

The Prelude

In marked contrast to this symbolic use of an animal are the autobiographical
recollections of beasts and birds recorded in *The Prelude.* Wordsworth's
childhood memories of animals are almost all those of a country boy to whom
animals are unexceptional: wild creatures are part of the landscape, while
domesticated beasts are a practical feature of a hard life. Sometimes the latter
may inspire affection by the quirky similarity of their behaviour to that of
human beings – like the favourite terrier which, when accompanying the young
Wordsworth on his walks, would come back and warn him whenever there was
someone else on the road, thus giving the budding poet (who was reciting his
verse out loud) time to 'hush his voice' and 'compose his gait' and so save him
from the rumour of being 'crazed in brain' (1805 *Prelude,* IV. 116–20). *The
Prelude,* and the other fragments of childhood narrative, trace the 'history of a
poet's mind' by starting with a boy who sees birds and beasts as objects, rather
than as feeling fellow-subjects: one who sets traps for woodcocks; robs birds'
nests; goes angling up the mountain brooks on rainy days; whips and spurs
his pony through Furness Abbey, and contemplates throwing a stone to break

---

[26] Letter to Wordsworth, April 1815: Lamb, *Letters,* III. 148. Byron reacted to the poem in
a similar way: 'I looked at Wordsworth's "milk-white Rylstone Doe," Hillo!' (letter to John
Murray, 25 March 1817, *BLJ,* V. 193).

[27] *The Letters of William and Dorothy Wordsworth: The Middle Years, part one,
1806–11,* ed by Ernest de Selincourt, 2nd edn, rev. by Mary Moorman (Oxford: Clarendon
Press, 1969) p. 222.

[28] Wordsworth wrote to Scott *(Poetical Works,* ed by de Selincourt and Darbishire, III.
542): 'I have followed (as I was in duty bound to do) the traditionary and common historic
records. Therefore I shall say in this case, a plague upon your industrious Antiquarianism, that
has put my fine story to confusion.'

up the image of the beautiful 'snow-white ram' reflected in the stream.[29] The same story is told in 'To a Butterfly' (1802):

> My sister Emmeline and I
> Together chased the butterfly!
> A very hunter I did rush
> Upon the prey; – with leaps and springs
> I followed thee from brake to bush;
> But she, God love her! feared to brush
> The dust from off its wings.
> (12–18)

The self-revealing honesty is deliberate: *The Prelude* is 'hard pastoral', and part of its purpose is realistically to portray a harsh country setting which has no time for sentimentality about either animals or human beings.[30] In Book V of the 1805 *Prelude* Wordsworth satirically draws at some length the contrasting picture of the model child of the educational theorists – a 'monster birth / Engendered by these too industrious times', a 'child, no child / But a dwarf man' (292–5) who, while 'The wandering beggars propagate his name,' and 'Dumb creatures find him tender as a nun' (303–4), is nevertheless presented as one to be pitied for his 'unnatural growth' (328).[31] The passage recalls the Wordsworth circle's dislike, described in Chapter Two, of the new children's books of the 1780s and 1790s which assumed the moral high ground on the subject of kindness to animals. Wordsworth accordingly makes a deliberate contrast between this theoretical infant prodigy and the 'race of real children' (436), such as himself:

> not too wise,
> Too learned, or too good; but wanton, fresh,
> And bandied up and down by love and hate.
> (436–8)

The implication is that, having been (healthily) little more than an animal himself as a child – 'a naked savage, in the thunder-shower' (1805 *Prelude,* I. 304) – Wordsworth has matured naturally and unforcedly into a man who can see animals as in need of his kindness. *The Prelude* follows through the history of the poet's attitude to animals to that of the late developer (in his early twenties) who 'would not strike a flower / As many a man would strike his horse,' and whose spirit was 'attuned' to 'gentleness of love' for animals.[32] *Home at Grasmere* shows how going away from home and coming back as an adult can transform the poet's view of animals to an observant benevolence. Now, at last, the 'inferior creatures' win from him 'those minute obeisances / Of tenderness' (1805 *Prelude,* VIII. 493), and these have become as important

---

[29] 1805 *Prelude:* I. 310–32; I. 333–50; VIII. 398–9; II. 123; *The Prelude,* MS Y(b) 15–26 in *The Prelude,* ed by Jonathan Wordsworth, p. 506.

[30] Bate, *Romantic Ecology,* p. 22: 'This is a working paradise ... the pastoral is hardened and differentiated from that of literary tradition.'

[31] Some of these phrases were considerably toned down in the 1850 *Prelude.*

[32] MS JJ *(The Prelude* ed by J Wordsworth and others, p. 493), and *1805 Prelude* VIII. 491–2.

to him as the 'first blessings' (VIII. 495), received direct from the hand of Nature.

Wordsworth's deployment of animals in his verse shows, then, a variety of ways of viewing them as cultural objects: from the country child's almost unthinking acceptance of them as part of the physical environment of 'hard pastoral', to a consciously benevolent and paternalistic attitude to them as 'inferior creatures', and a conception of them as spiritual or quasi-religious symbols. There is throughout his work a preoccupation with the problem of animals' status as marginal beings between nature and culture, and with their inability – or ability – to speak or express themselves, which highlights issues concerned with the poet's own powers and difficulties of articulation and his belief in a 'natural' or Adamic language beyond and more powerful than human words. The question that remains to be answered here is how far such a position can be measured against the characteristics of a modern 'ecological' viewpoint, and what congruence might be expected between attitudes separated by nearly two hundred years.

Ecological considerations

One proto-ecological theme which is often found in both the 'hard' and the 'soft' varieties of pastoral is conspicuous by its absence from Wordsworth's approaches to animals: that is, the omnipresence in nature of struggle and violent death, whether this is characterized as 'the survival of the fittest' or the 'Et in Arcadia ego' refrain. In the Romantic period, this can take the form of a meditation on the cruelty and bloodthirstiness of animals: an issue which was one of those which most concerned those Natural Theologians and others of this time who wished to see the ordering of the natural world as evidence of God's beneficence towards humankind. Erasmus Darwin's religious unorthodoxy finds expression in his insistence upon the way in which savagery is inherent to nature:

> The wolf, escorted by his milk-drawn dam,
> Unknown to mercy, tears the guiltless lamb;
> The towering eagle, darting from above,
> Unfeeling rends the inoffensive dove;
> The lamb and dove on living nature feed,
> Crop the young herb, or crush the embryon seed.
> (*Temple of Nature*, IV. 17–22)

This theme is also found, with an added intensity of human distress, in Keats's 'Epistle to J. H. Reynolds Esq.' (1819):

> I was at home
> And should have been most happy, – but I saw
> Too far into the sea, where every maw
> The greater on the less feeds evermore. –
> But I saw too distinct into the core
> Of an eternal fierce destruction,
> And so from happiness I far was gone.
> Still I am sick of it, and tho', to-day,

I've gather'd young spring-leaves, and flowers gay
Of periwinkle and wild strawberry,
Still do I that most fierce destruction see, –
The Shark at savage prey, – the Hawk at pounce, –
The gentle Robin, like a Pard or Ounce,
Ravening a worm.[33]
(92–105)

In its rare appearances in Wordsworth's writing this motif of bloodthirstiness is characterized by that obliquity of vision which has been noted earlier in his approaches to animals. In *Home at Grasmere*, for instance, whereas the earlier Manuscript B acknowledges (845–6) how Nature 'exacts / Her tribute of inevitable pain' (even in Grasmere), this passage and any other reference to nature's inherent violence is excised from the revised Manuscript D. A short poem of 1802, 'The Redbreast chasing the Butterfly', takes a resolutely childlike approach to what Tennyson was to characterize five decades later as 'Nature, red in tooth and claw / With ravine' *(In Memoriam,* stanza lvi)[34]. Wordsworth subjects the bird's natural predatory instincts to a rather mawkish sentimentality:

Can this be the bird, to man so good,
That, after their bewildering,
Covered with leaves the little children,
    So painfully in the wood?
What ailed thee, Robin, that thou couldst pursue
    A beautiful creature,
That is gentle by nature? ...
O pious Bird! whom man loves best,
Love him, or leave him alone!
(20–39)

In an 1805 poem, 'Fidelity', which celebrates the behaviour of a dog which remained with its dead master's body on the mountain for three months, the poet specifically questions how the animal was 'nourished here through such

---

[33] Book IV of *The Temple of Nature* (23–64) also lies behind this passage:
Nor spares the loud owl in her dusky flight,
Smit with sweet notes, the minstrel of the night;
Nor spares, enamour'd of his radiant form,
The hungry nightingale the glowing worm. ...
In ocean's pearly haunts, the wave beneath
Sits the grim monarch of insatiate Death;
The shark rapacious with descending blow
Darts on the scaly brood, that swims below;
The crawling crocodiles, beneath that move,
Arrest with rising jaw the tribes above;
With monstrous gape sepulchral whales devour
Shoals at a gulp, a million in an hour.
Air, earth, and ocean, to astonish'd day
One scene of blood, one mighty tomb display!
[34] *The Poems of Tennyson in Three Volumes: Second Edition Incorporating the Trinity College Manuscripts,* ed by Christopher Ricks (Harlow: Longman, 1969).

long time' (62), but closes his eyes to the strong possibility that the dog may have eaten from its master's corpse.[35] Such unwillingness to look steadily at this aspect of the reality of nature is characteristic, as I discuss in Chapter Six, of pre-Darwinian incapacity to relinquish a conception of the universe ordered according to human needs and in line with human perceptions of benevolence, beauty and truth: an inability to recognize, as Lawrence Buell has it, that 'the human interest is not understood to be the only legitimate interest.'[36]

Recent studies have shown that the Romantics were working in a period which had many – though not all – the concepts available for a 'modern' consideration of humankind as part of the natural environment.[37] Donald Worster and Jonathan Bate have claimed a kind of spiritual ecological consciousness *avant la lettre* for Wordsworth and other Romantic and Victorian poets. Worster, for instance considers (p. 58) that 'The Romantic approach to nature was fundamentally ecological: that is, it was concerned with relation, interdependence, and holism', and (p. 82) that

> at the very core of this Romantic view of nature was what later generations would come to call an ecological perspective: that is, a search for holistic or integrated perception, and emphasis of interdependence and relatedness in nature, and an intense desire to restore man to a place of intimate intercourse with the vast organism that constitutes the earth.

Bate (p. 40) sees that, while

> Scientists made it their business to describe the intricate economy of nature; Romantics made it theirs to teach human beings how to live as part of it ... The 'Romantic ecology' reverences the green earth because it recognizes that neither physically nor psychologically can we live without green things.

It is true that many of the ideas concerned with the seventeenth- and eighteenth-century concept of 'the economy of nature' are similar to those now brought together under the label of ecology: particularly the quality of relatedness and mutuality among living bodies. 'Natural Philosophy' is described, for example, in the Wordsworths' 1797 edition of the *Encyclopaedia Britannica* (XII. 670), as 'that art or science which considers the powers and properties of natural bodies, *and their mutual actions on one another,*' (my italics) and which 'recites the action of two or more bodies of the same or different kinds upon one another'. The *Encyclopaedia*'s entry on 'Natural History' (XII. 665) borrows a phrase (without acknowledgement) from Benjamin Stillingfleet's introduction to Linnaeus's essay 'The Oeconomy

---

[35] Some stanzas of this poem are inscribed on a monument on the top of Helvellyn in the Lake District, so many generations of tourists may also have closed their eyes to this possibility.

[36] Lawrence Buell, *The Environmental Imagination: Thoreau, Nature Writing, and the Formation of American Culture* (Cambridge, Mass.: Harvard University Press, 1995) p. 7.

[37] See Bate, *Romantic Ecology,* Bewell, Roe and Worster.

of Nature' (1749) which describes it as presenting 'as it were in a map, ... the several parts of nature, their connections and dependencies'.[38] Further, the *Encyclopaedia* uses Robert Boyle's words to describe 'Nature' (XII. 671) as 'the aggregate of the bodies that make up the world in its present state, considered as a principle; by virtue whereof they act and suffer, according to the laws of motion prescribeed by the Author of all things.'

But it is in the last few words of Boyle's sentence that the differences between 'economy of nature' and 'ecology' begin to emerge. The *Encyclopaedia* constantly reminds its readers that the works of nature are to be seen (and deliberately viewed) as the book of God, and that the world was made by God for the benefit of humankind. All these 'treasures of nature', claims the entry on 'Natural History' (XII. 664, tacitly quoting Linnaeus),

> so artfully contrived, so wonderfully propagated, so providentially supported throughout her three kingdoms, seem intended by the Creator for the sake of man. Every thing may be made subservient to his use, if not immediately, yet mediately; not so to that of other animals. By the help of reason, man tames the fiercest animals; pursues and catches the swiftest; nay, he is able to reach even those which lie hid in the bottom of the sea.

And, the entry concludes,

> In short, when we follow the series of created things, and consider how providently one is made for the sake of another, the matter comes to this, that all things are made for the sake of man; and for this end more especially, that he, by admiring the works of the Creator, should extol his glory, and at once enjoy all those things of which he stands in need, in order to pass his life conveniently and pleasantly.

Protestant theologians of the sixteenth and seventeenth centuries had been wont to cite the destruction of the perfection of Eden and the wildness of animals that came with the Fall to remind humanity of their fallen state, but this interpretation seems to have been laid aside by Natural Theology, which was concerned to demonstrate the beneficence of the world towards humankind.[39]

The *Encyclopaedia* entries represent the received wisdom of educated people of the period and not necessarily, of course, Wordsworth's own views. But statements such as the one in the 1814 Preface to *The Excursion* (63–8) do seem to indicate that – whatever Wordsworth's conception of the Deity – the notion of the centrality of humankind to the overarching order of the universe was not one from which he wished to distance himself:

---

[38] Carolus Linnaeus, 'The Oeconomy of Nature', in *Miscellaneous Tracts Relating to Natural History, Husbandry and Physick*, trans. (from *Amoenitates Academicae* by Carolus Linnaeus) by Benjamin Stillingfleet (London, 1749).

[39] See Erica Fudge, *Perceiving Animals: Humans and Beasts in Early Modern English Culture* (Basingstoke: Macmillan, 2000) pp. 37–8, citing *A Commentarie of John Caluine, Vpon the first booke of Moses called* Genesis (1578): 'so soone as he [Adam] began to be obstinate and rebellious against God, he felt the fierceness of the brute beastes against him.'

How exquisitely the individual Mind
(And the progressive powers perhaps no less
Of the whole species) to the external World
Is fitted – how exquisitely too –
Theme this but little heard of among men –
The external world is fitted to the Mind.

Indeed, given the prevalence of this theme in works such as the *Encyclopaedia Britannica,* the only surprise in this statement is Wordsworth's claim that the second part of it is 'but little heard of among men'.

Wordsworth's principles – his chosen way of living, his concern for the protection of the geology and the way of life of the Lake District, his vision of the inter-dependencies between man and nature in the Vale of Grasmere, and his expressed gentleness, tenderness and 'minute obeisances' to beasts and birds – may have affinities with the behaviour of those who, at the beginning of the twenty-first century, are concerned about ecology and environmentalism. But his poetic practice, in terms of his approach to animals in his verse, is less amenable to an ecological interpretation. Here, his attitude to the natural environment must be seen in the light of his express exclusion of 'the inferior creatures' from the powerful influence of 'Nature' upon his writing, his unwillingness to confront the reality of the struggle for existence (an essential feature of ecology) among beasts and birds, and his adherence to ideas about the centrality of humankind in a divinely-ordered universe. Tony Pinkney's recent consideration of 'Romantic Ecology' describes ecological criticism as that which requires an art-work to approach the natural realm by celebrating plant, birds, and animals for themselves, rather than subjugating them to human purposes, however benign these may seem.[40] Concentrating on the specific area of animals does throw into relief issues concerned with the meeting-ground of literature and ecology, because animals are themselves simultaneously both cultural and natural. It also helps to remind us that 'green' politics cannot be confined to things that are, literally, green. Pinkney's examples of 'how *not* to do it' include (p. 414) Shelley's 'To a Skylark': 'Hail to thee, blithe Spirit! / Bird thou never wert', where

> the skylark's own identity, its actual autonomous being in its indigenous environment, is at once cancelled out. The bird instead is turned into a symbol, into a representative of the poet's own inner aspirations; it is folded back into the subjectivity of the author rather than respected in its objectivity and otherness.

Pinkney supplements the example from Shelley with Wordsworth's description, in 'To a Butterfly', of the insect as 'Historian of my Infancy', and he might also have added Wordsworth's well-known evocation of the cuckoo: 'Shall I call thee bird, or but a wandering voice?' As has been noted in this chapter, Wordsworth's general ease in bestowing a voice upon inanimate, physical features, such as mountains, woods and rivers, is matched by the

---

[40] Tony Pinkney, 'Romantic Ecology' in *A Companion to Romanticism* ed by Duncan Wu (Oxford: Blackwell, 1998) pp. 411–19 (p. 414).

complexities of the literary artifice he needs to deploy to interpret the expression of animate nonhuman creatures.

As Marjorie Levinson has remarked *(Wordsworth,* p. 105), 'to define oneself as a seer "*into* the life of things"' may be 'to forfeit the *life* of things', and the meeting of Wordsworth's 'egotistical sublime' with the radically different egotism of beasts and birds sets up tensions which make the poet's handling of this area a fruitful one to study. Although, however, it is unlikely that modern readers will succeed in finding exemplary ecological or proto-ecological attitudes in the approaches to animals in Wordsworth's verse, this does not necessarily mean that his interest in nature must always be suspect on political grounds – used simply as a cloak for disguising a change of ideological allegiance. The recent extensive debate about other aspects of Romantic nature has helped us to appreciate the deep historicism of our own perspectives on the natural world, and another alternative for late-twentieth century readers with an ecological perspective means the application of further study to elucidate the full range of preoccupations Wordsworth had in this area, together with a willingness to respect the differences as well as the similarities of perception we find between his approach and our own.

Chapter Six

Evolutionary Animals: Science and Imagination Between the Darwins

If the labours of men of Science should ever create any material revolution, direct or indirect, in our condition, and in the impressions which we habitually receive, the Poet will sleep then no more than at present, but he will be ready to follow the steps of the man of Science, not only in those general indirect effects, but he will be at his side, carrying sensation into the midst of the objects of the Science itself. The remotest discoveries of the Chemist, the Botanist, or Mineralogist, will be as proper objects of the Poet's art as any upon which it can be employed. ... If the time should ever come when what is now called Science, thus familiarized to men, shall be ready to put on, as it were, a form of flesh and blood, the Poet will lend his divine spirit to aid the transfiguration and will welcome the Being thus produced, as a dear and genuine inmate of the household of man. (Lines added in 1802 to the *Preface* to *Lyrical Ballads,* 1800)

Wordsworth's description of the relationship between 'men of Science' and Poets surprises not so much by its recommendation that they should work together, as by the accuracy of its prophecy about the extent to which they would develop apart. At the time that Wordsworth was writing, the studies of the natural and the human world were not widely separated, and indeed the turn of the century saw a vogue for what Coleridge called 'darwinizing': the kind of intelligent but sometimes ungrounded theoretical speculation on natural phenomena in verse made famous by Dr Erasmus Darwin in *The Botanic Garden* (1791) and *The Temple of Nature* (1803).[1] Writers such as Dr Darwin, and before him Richard Payne Knight, Mark Akenside and James Thomson, had regarded it as a relatively unproblematic part of their life's work to enable 'science' to 'put on a form of flesh and blood' through poetry, and it was in the same preface that he made these remarks that Wordsworth also criticized the hackneyed overuse of stylistic devices such as personification which this process seemed to involve.[2]

---

[1] 'Notes on Stillingfleet by S. T. Coleridge', *The Athenaeum*, 2474, (1875) p. 423. See Peter Graham, 'Byronic Darwinizing' in *Lord Byron: A Multidisciplinary Open Forum, proceedings of the 23rd International Byron Conference, Versailles, 26–30 June, 1997,* ed by Thérèse Tessier (n. p., 1999) pp. 125–34.

[2] Knight, *The Progress of Civil Society* (1796); Akenside, *Pleasures of Imagination* (1744 and 1757) and Thomson, *The Seasons* (1726–30). *Preface* to *Lyrical Ballads,* 1800: 'Except in a very few instances the Reader will find no personifications of abstract ideas in these volumes.' Wordsworth may have been reacting to a lecture by Darwin's friend Joseph Priestley, who claimed it was 'of prodigious advantage in treating of inanimate things, or merely of brute animals, to introduce frequent allusions to human actions and sentiments,

In 1802 the word 'science' still retained the more general sense of any knowledge acquired by systemized study, although Wordsworth's examples are of areas which have come to be regarded as unequivocally 'scientific' in the modern sense. The terms still in common use in Wordsworth's period for the observation and analysis of the natural world were 'Natural History' and 'Natural Philosophy'.[3] A generation later, the concept of 'science' was still in the course of developing its more specialist meaning, when in 1833 at the third annual meeting of the British Association for the Advancement of Science Coleridge discussed the adoption of the word 'scientist'.[4] And even in the mid-nineteenth century, as Gillian Beer points out, scientists still shared a common language with non-scientific educated readers and writers of their time:

> Their texts could be read very much as literary texts. Moreover, scientists themselves in their texts drew openly upon literary, historical and philosophical material as part of their arguments. ... Because of the shared discourse not only *ideas* but metaphors, myths and narrative patterns could move rapidly and freely to and fro between scientists and non-scientists: though not without frequent creative misprision.[5]

Scientists of this period such as Humphrey Davy and James Clerk Maxwell wrote verse about their scientific discoveries without embarrassment, and Michael Faraday spoke in 1845 of how he was

> struggling to exert my poetical ideas just now for the discovery of analogies and remote figures respecting the earth, sun, and all sorts of things – for I think that is the true way (corrected by judgment) to work out a discovery.[6]

Wordsworth's stance has perhaps contributed to what Hermione de Almeida calls 'a common but otherwise insupportable twentieth-century assumption of the alienation of Romanticism from science', although her own study of Keats and medicine, John Wyatt's of Wordsworth and the geologists, Carl Grabo's of Shelley and science, and Levere's of Coleridge and science have shown the

---

where any resemblance will make it natural. This converts everything we treat of into thinking and acting beings. We see life, sense, intelligence everywhere.' *A Course of Lectures in Oratory and Criticism* (1777), quoted by Maureen McNeil, *Under the Banner of Science: Erasmus Darwin and his Age* (Manchester: Manchester University Press) p. 198.

[3] See *Encyclopaedia Britannica*, 3rd edn, 1797, XII. 663 and 670.

[4] See William Whewell's account in the *Quarterly Review* LI (1834) p. 59: *'Philosophers* was felt to be too wide and too lofty a term, and was very properly forbidden them by Mr Coleridge, both in his capacity as philologer and metaphysician; *savans* [sic] was rather assuming, besides being French instead of English; some ingenious gentleman proposed that, by analogy with *artist,* they might form *scientist,* and added that there could be no scruple in making free with this termination, when we have such words as *sciolist, economist,* and *atheist* – but it was not generally palatable.'

[5] Gillian Beer, *Darwin's Plots: Evolutionary Narrative in Darwin, George Eliot and Nineteenth-Century Fiction* (London: Routledge & Kegan Paul, 1983) pp. 6–7.

[6] Quoted in J. A. V. Chapple, *Science and Literature in the Nineteenth Century* (Basingstoke: Macmillan, 1986) p. 38.

extent to which some Romantic poets used and understood scientific concepts.[7] Although Erasmus Darwin's direct manner of teaching science through verse did not continue in the nineteenth century (except in the form of literature for children) there is ample evidence of a more complex symbiosis between scientific and imaginative thinking which developed during the half-century after his death.

This chapter considers this developing situation of differentiation and competition between 'scientists' and poets in the Romantic period in the light of a specific debate which was to dominate both science and literature (as well as philosophy and theology) in the mid-century – that of the evolution of biological species. In the Romantic period the topic goes by a variety of terms: expressions such as transformism, metamorphosis, parallelism, transmutation, progression, development and perfectibility reflect the richness of pre-Darwinian hypothesizing on this subject and also the lack of consensus in the period about the *mechanism* by which species changed and became extinct. It is only with Darwin's explicit title in 1859 (*On the Origin of Species by Means of Natural Selection; or, The Preservation of Favoured Races in the Struggle for Life*) that the question of causes, means and outcomes is asked, let alone answered, in recognizably modern terms. The concept of evolution itself can, however, be traced back to classical authors, including Horace, Cicero and Lucretius; it had been increasingly 'in the air' since fossil bones had been identified by authors such as Buffon in the eighteenth century as being those of species no longer in existence, and it was in terms of this topic that various urgent questions about the relationship between human beings and other species began to coalesce in the late eighteenth and early nineteenth centuries.[8]

My study of the expression of early evolutionary ideas explores Romantic fictions about evolutionary themes by taking the notion of 'species' as central, and paying particular attention to issues of kinship, miscegenation, and monstrosity: the particular science of which ('teratology') was one of those to

---

[7] De Almeida, *Romantic Medicine and John Keats* (Oxford: Oxford University Press, 1991) p. 3; John Wyatt, *Wordsworth and the Geologists* (Cambridge: Cambridge University Press, 1995); Carl Grabo, *A Newton among Poets: Shelley's Use of Science in* Prometheus Unbound (Chapel Hill: University of North Carolina Press, 1930).

[8] See Thomas, p. 130: 'In classical antiquity, Protagoras, Diodorus Siculus, Lucretius, Horace, Cicero and Vitruvius had all suggested that man had made only a gradual ascent from a bestial condition, developing language and civilisation over a long period of time', and Buffon in vol. IV of his *Histoire naturelle* (1753): 'If we regard the matter thus, not only the ass and the horse, but even man himself, the apes, the quadrupeds, and all animals might be regarded but as forming members of one and the same family ... we might be driven to admit that the ape is of the family of man, that he is but a degenerate man, and that he and man have a common ancestor, even as the ass and horse have had. It would follow then that every family, whether animal or vegetable, had sprung from a single stock, which after a succession of generations, had become higher in the case of some of its descendants and lower in that of others.' Fearful of being branded a religious sceptic, Buffon then added: 'But no! It is certain from revelation that all animals have alike been favoured with the grace of an act of direct creation, and that the first pair of every species issued full formed from the hands of the Creator.' *Histoire naturelle, générale et particulière*, 44 vols (Paris: Imprimerie Royale, puis Plassan, 1749–1804) IV. 381, translated here by Samuel Butler in *Evolution, Old and New; Or, The Theories of Buffon, Dr Erasmus Darwin, and Lamarck, as Compared with that of Charles Darwin* (London: Fifield, 1911) p. 88.

be separated out and given its own name in the early nineteenth century.[9] In common with the practice of the time, this exploration brings together theological and philosophical ideas with biological and geological ones, in the context of poetic writing. It demonstrates the continuity of the debate beyond the strictly 'Romantic' period by using examples from Tennyson, but it concentrates on Keats (whose approach to evolutionary themes has been studied by de Almeida in the context of his medical knowledge, but without any specific attention to animals) and on Byron. Byron's work has not, like that of many of his Romantic contemporaries, been subjected to a full-scale analysis of its scientific content. Perhaps because of his own expressed dislike of 'system' (following Buffon) and his wish as an aristocrat not to be seen to labour too distinctly at anything, commentators have continued to regard Byron's interest in metaphysics as superficial.[10] The geological theory of Catastrophism is described by Marilyn Gaull as one which

> appealed to the post-war generation of emotionally over-wrought sensation-seekers who were supporting Gothic melodrama at Covent Garden, who made Byron a hero, who purchased a record number of broadsides recounting gruesome murders, and who genuinely believed in the apocalyptic yearnings of Joanna Southcott.[11]

Byron's own deployment of Catastrophism has been characterized as no more than 'the cultivated man's polite interest in the peculiar geological theory that was midway between the Book of Genesis and the book of Charles Darwin', while the metaphysical drama, *Cain,* in which he made use of the theory as part of the plot, is dismissed as 'as potent an affirmation of Byron's bankruptcy as a philosophical poet as we are likely to find. ... bad poetry and worse drama'.[12] I shall suggest here that in fact Byron and Keats were particularly well aware of and knowledgeable about the debate about 'science' and 'poetry' in this period, and that these late Romantic writings deliberately bring to what Ludmilla Jordanova has called the general 'Romantic preoccupation with Creation myths' a self-consciousness about the deployment of 'scientific' themes in imaginative literature and a critique of the way in which 'scientific' analogies were being developed at this time.[13]

---

[9] By Isidore Geoffroy Saint-Hilaire in 1830 *(Histoire générale et particulière des anomalies de l'organisation chez l'homme et les animaux ... ou, traité de tératologie,* 3 vols (Paris, 1832–37).

[10] Letter to Thomas Moore, 1 June 1818 *(BLJ,* VI. 46–7) 'when a man talks of system, his case is hopeless'; and, 'I thought that poetry was an *art,* or an *attribute,* and not a *profession.'* Buffon wrote at the beginning of the *Histoire naturelle:* 'they build systems on uncertain facts, which have never been examined, and which only serve to show the tendency of man ... to find regularity where variety reigns.'

[11] Marilyn Gaull, *English Romanticism: The Human Context* (New York: Norton, 1988) p. 212.

[12] Lionel Stevenson, *Darwin among the Poets,* 1932 (repr. New York: Russell & Russell, 1963) p. 34, and Philip W. Martin, *Byron: A Poet before his Public* (Cambridge: Cambridge University Press, 1982) p. 148.

[13] Ludmilla Jordanova, 'Nature's Powers: A Reading of Lamarck's Distinction between Creation and Production' in *History, Humanity and Evolution, Essays for John C. Greene,* ed by James R. Moore (Cambridge: Cambridge University Press, 1989) 71–98 (p. 88).

As Charles Darwin's editor J. W. Burrow notes, by the time *The Origin* was published in 1859, 'the theory of evolution in biology was already an old, even a discredited one. Darwin, in later editions of *The Origin,* listed over thirty predecessors and was still accused of lack of generosity.'[14] Many of the materials mentioned by Darwin in his 'Historical Sketch' to *The Origin* were known to Romantic-period writers, including classical texts such as Aristotle's; widely-read eighteenth-century authors such as Buffon; Goethe; W. C. Wells (a physician of St Thomas' Hospital, whose work on natural selection according to 'beauty' is believed to have been familiar to Keats) and Charles's grandfather Erasmus Darwin.[15] Other authors mentioned by Darwin as among his sources for *The Origin* include Baron Cuvier (one of Byron's major sources for *Cain)* and Cuvier's colleague and rival, the Chevalier de Lamarck. Lamarck's theories proposed that new characteristics which were acquired by animals in response to changing conditions (such as giraffes developing long necks through stretching to reach high leafy branches) could be passed on to their offspring, and these ideas would have been known to Byron from his reading of Cuvier, if not directly.[16] A potent influence on Charles Darwin was Thomas Malthus's essay 'On the Principle of Population' (1798), frequently alluded to by Byron, which posited that the human population would quickly outstrip food production, because population multiplied by geometrical ratio while food production could be made to increase only arithmetically.[17] Darwin described his own theory about the struggle for existence as 'the doctrine of Malthus applied with manifold force to the whole animal and vegetable kingdoms' *(The Origin,* p. 117). Another important source shared by Byron and Darwin, which is explored further below, lies in what I believe is their common interrogation of William Paley's influential argument, expressed in his *Natural Theology* (1802), that the beneficence of the deity towards humankind was specifically demonstrated by the design of the natural world.

Animals from the inside

The conception of the 'poetical character' which Keats gave to Richard Woodhouse in October 1818 throws light upon an important aspect of his approaches to animals:

---

[14] *The Origin of Species,* 1st edn, 1859, ed by J. W. Burrow (Harmondsworth: Penguin, 1968) p. 27.

[15] For examples of Byron's quotations of Buffon see *LBCMP* p. 126 and *LBCMP* p. 167. Keats read Buffon in Barr's 1792 translation in the spring of 1818 while nursing his brother Tom: see *The Letters of John Keats 1814–1821,* ed by Hyder Edward Rollins, 2 vols (Cambridge Mass.: Harvard University Press, 1958) I. 255n and I. 263n. Byron satirized Erasmus Darwin as 'the mighty master of unmeaning rhyme' in *English Bards and Scotch Reviewers,* 891–902. Lady Byron recorded him reading Darwin's *Zoonomia* in 1815 (Lovell, *His Very Self,* p. 105).

[16] See Lamarck's *Philosophie zoologique,* 2 vols (Paris, 1809) and *Histoire naturelle des animaux sans vertèbres,* 7 vols (Paris, 1815–22).

[17] For Byron on Malthus see, for example *Don Juan,* I. 1041–8; VI. 151–2; XI. 235–9; XII. 108–10; XV. 289–304. Byron was acquainted with Malthus personally, see *BLJ,* IX. 19.

[A]s to the poetical character itself (I mean that sort of which, if I am anything, I am a member – that sort distinguished from the Wordsworthian or egotistical sublime, which is a thing per se and stands alone), it is not itself – it has no self – it is everything and nothing – it has no character – it enjoys light and shade – it lives in gusto, be it foul or fair, high or low, rich or poor, mean or elevated. It has as much delight in conceiving and Iago as an Imogen. What shocks the virtuous philosopher delights the chameleon poet. ... A poet is the most unpoetical of any thing in existence, because he has no identity, he is continually in for – and filling – some other body. ... When I am in a room with People if ever I am free from speculating on creations of my own brain, then not myself goes home to myself: but the identity of every one in the room begins to press upon me that, I am in a very little time annihilated.

In an earlier letter, Keats specifically associates this 'filling some other body' with a propensity – indeed, an involuntary obligation – to engage imaginatively with animals at the most basic physical level: '[I]f a Sparrow come before my Window,' he tells Benjamin Bailey (*Letters* 1. 186), 'I take part in its existence and pick about the Gravel'. The 'Ode to Nightingale' recreates another similar moment when the poet almost 'becomes' the bird, finding himself 'too happy in thine happiness' – until the word 'forlorn' 'toll[s]' him 'back from thee to my sole self'. Such an engagement, and a corresponding ability to articulate in sensuous terms what it feels like to be re-embodied in an animal, can be traced throughout his poems: in, for example, the hare which 'limped trembling through the frozen grass'; in the bees who 'think warm days will never cease / For summer has o'er-brimmed their clammy cells', and particularly in the minnows

> Staying their wavy bodies 'gainst the streams,
> To taste the luxury of sunny beams
> Tempered with coolness. How they ever wrestle
> With their own sweet delight, and ever nestle
> Their silvery bellies on the pebbly sand.[18]

It is, however, another aspect of Keats's internalisation of animals – another way in which he enters into their nature – that I want to explore in this study. Here I want to look at how he approaches animals from the *anatomical* inside, in his capacity as a student of science, aware of the theories and practices of his time which anatomised, dissected and displayed them inside-out. Keats, like both Darwins, had a professional medical education, and he encountered evolutionary theories during his year's attendance and study at the United Hospitals of Guy's and St Thomas' which enabled him to be licensed as an apothecary and general practitioner in 1816.[19] Providing a link between the German *Naturphilosophie* ideas associated with Coleridge and the theories of biological development to which Keats was exposed was Joseph Henry Green,

---

[18] 'Eve of St Agnes', 3; 'Ode to Autumn', 9–11; 'I stood tip-toe upon a little hill', 73–7.

[19] Having been apprenticed for five and a half years to Thomas Hammond, apothecary, in Edmonton, Keats was registered at Guy's during the first year in which new regulations were in place for the more rigorous training of apothecaries.

friend and disciple of Coleridge, who was Keats's teacher of morbid anatomy and dissection. It was Green's thesis (following Lorenz Oken's *Abriss der Naturphilosophie* ('Outline of Nature Philosophy', 1805) and *Die Zeugung* ('Generation', 1805) that the development of the human embryo followed the order of living beings envisioned in the 'Great Chain of Being' – from microscopic creatures, through reptiles, birds and mammals up to man – and that therefore a premature human foetus, or deformed birth, represented the 'lower' form showing through in the higher:

> Congenital Defects, hitherto comprehended under the vague definition of monstrosity, are to be explained by the development of the embryo being interrupted and arrested at some early stage of its regular evolution, and the defective form, which is the result, is analogous to the form and structure of an inferior class. And thus if in the human embryo these defective forms constitute a series of transient epochs, which are repetitious of the types, that denote the ascending scale of animated being, in like manner all the lower forms in relation to the highest may be regarded as abortions, by anticipation of nature's mature work, the human frame.[20]

De Almeida (p. 267) notes how, in the operating theatres of hospitals such as Guy's in Keats's time (and still, in the Gordon Museum at Guy's) jars were displayed filled with freak embryos and monstrosities. Charles Darwin took from such studies the idea that the similarity of embryos of different species demonstrated their common origin, but not the concomitant theory (current in various forms throughout the early nineteenth century) that such development demonstrated the superiority of humankind over other species, and that other species were in effect 'trials' for humankind, which was itself in the course of progressive development towards perfection or a 'higher type'. In this form it was a source for Tennyson's *The Palace of Art* (1832):

> 'From change to change four times within the womb
>   The brain is moulded,' she began,
> 'So through all phases of all thought I come
>   Into the perfect man.
>
> 'All nature widens upward: evermore
>   The simpler essence lower lies,
> More complex is more perfect, owning more
>   Discourse, more widely wise.'
> (*The Palace of Art,* 128, additional lines 13–20)[21]

---

[20] Joseph Henry Green, *Vital Dynamics: The Hunterian Oration Before the Royal College of Surgeons in London, 14 February 1840* (London: Pickering, 1840) p. 40.

[21] When republishing in 1842, by which time ontological transmutation from 'lower' forms of life to human had been shown to be figurative rather than literal, Tennyson changed these lines to '"From shape to shape at first within the womb / The brain is modell'd "'. See John Killham, *Tennyson and The Princess: Reflections of an Age* (London: Athlone Press, 1958) pp. 242–3.

There were also political consequences to be drawn from such a thesis, and Adrian Desmond has shown how between the second and fifth decades of the nineteenth century the evolutionary debate was polarized between British radical and conservative political commentators.[22] Lamarckian ideas, which held out the possibility that advantages learned by animals might be passed on to their offspring, were welcomed by radicals because they provided a mechanism for the stabilization of mental and physical (human) attributes derived from education (see Desmond, 'Artisan Resistance', p. 94); but they were execrated by anti-democrats like Green since he emphatically denied any 'power in the lower to become, or to assume the rank and privilege of, the higher'. Were this not so, 'cause' in nature would always be 'meaner and feebler than the effect', making 'blindness the source of light' and 'ignorance ... the parent of mind' (Green, cited by Desmond in 'Lamarckism and Democracy', p. 122). Coleridge similarly dismissed Lamarckian evolution as an 'Ouran Outang theology of the origin of the human race' and claimed that the 'plebification' of science in 'lecture-bazaars under the absurd name of universities' bred only further blasphemy and discontent among the working classes.[23] Keats, on the other hand, who was described by George Felton Mathew as 'of the skeptical and republican school ... an advocate of the innovations which were making progress in his time ... a faultfinder with everything established', evidently saw the possibilities of Lamarckian evolution and extinction in another light.[24] 'Parsons', he wrote in 1819,

> will always keep their Character, but as it is said there are some animals, the Ancients knew, which we do not; let us hope our posterity will not miss the black badger with tri-cornered hat. Who knows but some reviser of Buffon or Pliny, may put an account of the Parson in the Appendix; No one will then beleive [sic] it any more than we beleive in the Phoenix. I think we may class the lawyer in the same natural history of Monsters. (Keats, *Letters*, II. 70)

The possibility of 'perfectibility' is passionately embraced by Tennyson in the Epilogue to *In Memoriam*, stanzas cxxviii–cxxx, through the mourner's hope that transmutation could lead to Man's evolution into 'the crowning race'. It is, however, flatly rejected by Keats:

> But in truth I do not at all believe in this sort of perfectibility – the nature of the world will not admit of it – the inhabitants of the world will correspond to itself – Let the fish philosophize the ice away from the Rivers in winter time and they shall be at continual play in the tepid delight of summer. ... The point at which Man may arrive is as far as the

---

[22] 'Artisan Resistance and Evolution in Britain, 1819–1848' in *Osiris* 2nd series, III (1987) pp. 77–110, and 'Lamarckism and Democracy: Corporations, Corruption and Comparative Anatomy in the 1830s' in James R. Moore (ed), pp. 99–130.

[23] *On the Constitution of the Church and State*, ed by John Colmer, *The Collected Works of Samuel Taylor Coleridge*, Bollingen Series (London: Routledge and Kegan Paul, 1976) pp. 66 and 69. The reference seems to be to University College London.

[24] Quoted in *The Keats Circle* ed by Hyder Edward Rollins, 2nd edn, 2 vols (Cambridge, Mass.: Harvard University Press, 1965) II. 185–6.

the paralel [sic] state in inanimate nature and no further – For instance
suppose a rose to have a sensation, it blooms on a beautiful morning it
enjoys itself – but there comes a cold wind, a hot sun – it cannot escape
it, it cannot destroy its annoyances – they are as native to the world as
itself: no more can man be happy in spite, the worldy [sic] elements
will prey upon his nature. (Keats, *Letters*, II. 101)

Keats's response to the embryological aspects of the theory of progression
(those which he learnt of through his medical training) is particularly evident in
his (metaphorical) descriptions of birth and metamorphosis. Barry Gradman
claims that 'Metamorphosis ... is a primary impulse of Keats's art, one of his
principal means of articulation' and contends that transformations usually have
positive results in Keats's poems. In fact, however, as Gradman's own account
demonstrates, there are numerous counter-instances where the transition from
one state to another seems to be associated with troubling sexual aspects, with
a disgust of fleshly animality, and with the possibility that a monstrous, mixed
form may emerge in place of the fully human one.[25] Sometimes
transformations which Keats himself claims to be beneficial are actually
ambiguous or deleterious; as in, for instance, a letter of 11 July 1819 to J. H.
Reynolds:

I have of late been moulting: not for fresh feathers & wings: they are
gone, and in their stead I hope to have a pair of patient sublunary legs. I
have altered, not from a Chrysalis into a butterfly, but the Contrary,
having two little loopholes, whence I may look out into the stage of the
world. (Keats, *Letters*, II. 128)

Circe's transformation of her human lovers into animals, as narrated by
Glaucus in Book III of *Endymion,* provides a grotesque and Gothic contrast to
the dreamy tone of the rest of the poem. Here men are trapped within animal
bodies, but agonisingly (as so often in Ovid's *Metamorphoses*[26]) retain their
human consciousness:

'And all around her shapes, wizard and brute,
Laughing, and wailing, grovelling, serpenting,
Shewing tooth, tusk, and venom-fang, and sting!
Oh, such deformities! ...
    'Remorseless as an infant's bier
She whisked against their eyes the sooty oil.
Whereat was heard a noise of painful toil,
Increasing gradual to a tempest rage,
Shrieks, yells, and groans of torture-pilgrimage,
Until their grievèd bodies 'gan to bloat
And puff from the tail's end to stifled throat.'
(Endymion, III. 500–3, 520–6)

---

[25] Barry Gradman, *Metamorphosis in Keats* (Brighton: Harvester Press, 1980) p. xv.
[26] See, for example, Actaeon, after his transformation into a deer by Diana in book III.

The odd simile of the 'remorseless ... infant's bier' and the idea of a 'torture-pilgrimage' associate this scene with a failed birth and with the painful progress of the soul through life to death, and this vision of Circe reverses the ideas of 'upward' evolutionary transition, which Keats had been taught during his medical training, by making the human state regress to the animal one. In the passage which follows, ancient mythological creatures emerge 'from the dark', whose hybrid human/animal nature is evident in their form and in the sexual licence of their behaviour: 'waggish fauns, and nymphs, and satyrs stark ... / Swifter than centaurs after rapine bent.' A man in the shape of an elephant begs to be

> 'delivered from this cumbrous flesh,
> From this gross, detestable, filthy mesh,
> And merely given to the cold, bleak air,'
> *(Endymion,* III. 551–3)

in an image which seems to recall the fossilized or frozen bodies of mammoths, as disinterred by palaeontologists.[27] The passage as a whole echoes the self-disgust of his own flesh and animality recorded (with a precision of medical terminology) in marginalia Keats wrote in a copy of Burton's *Anatomy of Melancholy:*

> Here is the old plague spot: the pestilence, the raw scrofula. I mean that there is nothing disgraces me in my own eyes so much as being one in a race of eyes, nose and mouth beings in a planet called the earth who all from Plato to Wesley have always mingled goatish, winnyish, lustful love with the abstract adoration of the deity.[28]

The Titans of *Hyperion* are of course themselves a race or species which is about to become extinct. H. W. Piper argued that Keats might have gleaned ideas for Oceanus's pronouncement of 'the eternal law / That first in beauty should be first in might' (II. 228–9) from a paper on natural selection according to 'beauty' by W. C. Wells, a physician at St Thomas' Hospital during Keats's period of training.[29] Wells's paper suggested that there was an evolution in races from the Negro to the white European: from ugliness to beauty, according to the supposed perception of beauty from the European point of view.[30] Oceanus prophesies that 'by that law, another race may drive / Our conquerors to mourn as we do now' (II. 228–31), and it is tempting to read this and the apparent equation of 'truth' and 'beauty' in the 'Ode on a Grecian Urn' (49–50) as part of an evolutionary racial theory of 'survival of the fittest'.

---

[27] Giovanni-Batista Gelli's *Circe* (1549) makes the elephant the only one of Circe's changelings to wish to be turned back into a man.

[28] Quoted in Hillas Smith, *Keats and Medicine* (Newport: Cross Publishing, 1995) p. 87.

[29] De Almeida, p. 261, quoting from 'Keats and W. C. Wells' by H. W. Piper, in *Review of English Studies* 25 (1949) pp. 158–9.

[30] One of Byron's iconoclastic touches in *Heaven and Earth* is to make the descendants of Cain (traditionally, the African races) greater in 'their stature and their beauty, / Their courage, strength and length of days' than the descendants of Seth, the son born to Adam in his old age and supposed to be the ancestor of the Europeans. See also *Don Juan,* XII. 559–61: 'But if I had been at Timbuctoo, there / No doubt I should be told that black is fair. / It is.'

There is no hint that Keats's Titans are Negroid or non-European, and in fact Fiona Stafford has convincingly made the case that Keats's comparison of them to 'a dismal cirque/of Druid stones, upon a forlorn moor' (II. 34–5) and his bestowing on Saturn 'Druid locks' (I. 136) indicate that he associated the Titans with the Celts, in the manner of Ossian.[31] The idea of a succession of superhuman races in *Hyperion* may nevertheless reflect Keats's understanding of the transference of supremacy in human history, and the replacement of one people (as much as one species) with another which is 'superior', more 'mighty', and therefore more beautiful.

Apollo's swift transition from humanity to godhead in *Hyperion* does not at first sight seem to be 'evolutionary', at least in Darwinian terms. The scene may, however, owe something to Cuvier's theories on Catastrophism, which posited that the animal species known from fossils had been made extinct by calamities which had overturned the entire outer crust of the globe. The 'knowledge enormous' which 'makes a god' of Apollo includes 'dire events, rebellions, / Majesties, sovran voices, agonies, / Creations and destroyings' (III. 114–16) which 'all at once / Pour into the wide hollows of [his] brain', and which are reminiscent of Cuvier's claim that 'nature also has had her intestine wars, and ... the surface of the globe has been much convulsed by successive revolutions and various catastrophes'.[32] That an 'off-stage' geological catastrophe, with the sudden transposition of water and dry land, may have been responsible for the Titans' fall is also suggested by the harsh features of the landscape where the defeated giants meet:

> a den where no insulting light
> Could glimmer on their tears; where their own groans
> They felt, but heard not, for the solid roar
> Of thunderous waterfalls and torrents hoarse,
> Pouring a constant bulk, uncertain where.
> Crag jutting forth to crag, and rocks that seemed
> Ever as if just rising from a sleep,
> Forehead to forehead joined their monstrous horns. ...
> Instead of thrones, hard flint they sat upon,
> Couches of rugged stone, and slaty ridge
> Stubborned with iron.
> *(Hyperion,* II. 5–17)

Sudden transformations like that of Apollo are also a key feature of Keats's *Lamia*. The classical Lamian metamorphosis is also, however, of a different kind from Darwinian evolutionary development, as Beer's distinction *(Darwin's Plots,* p. 111) makes clear: '"Omnia mutantur, nihil interit." Everything changes, nothing dies. Ovid's assertion in *Metamorphoses* marks one crucial distinction between the idea of metamorphosis and Darwin's theory of evolution. Darwin's theory required extinction.' Similarly, it is of a

---

[31] Fiona Stafford, 'Fingal and the Fallen Angels: Macpherson, Milton and Romantic Titanism': paper read at 'Influence and Intertextuality' conference, University of Bristol, 24 May 1997.

[32] *Essay of the Theory of the Earth translated from the French of M Cuvier ... by Robert Kerr ... with Mineralogical Notes and an Account of Cuvier's Geological Discoveries by Professor Jameson* (Edinburgh: William Blackwood and John Murray, 1813) p. 7.

neo-Platonic metamorphosis, not an evolutionary process, that Shelley is thinking when he makes the spirit of 'The Cloud' declare: 'I change but I cannot die' (76). Unlike her human lover Lycius, Lamia does not die but 'vanishes' at the end of the poem: apparently her hard-held humanity has given way under Apollonius's scientific glare and she has reverted to a serpent. But although she 'breathe[s] death-breath' (II. 299), we know that she has the capacity to 'fade at self-will' (II. 142): she claims she has been changed previously from woman to snake; we have seen her change from snake to woman, and there is no reason to think that she has ceased to exist as a result of this latest transformation.

As John Hedley-Brooke (p. 45) has pointed out: 'For Darwin the fossil record, imperfect though it was, disclosed the details of a process which, starting from a single animate source, had occurred in time once and for all. The history of life was a unique history.' Early-nineteenth-century ideas of evolution were even more firmly characterized as 'forward-only' since, although Darwin's theory allows for the possibility that species may stay the same if the environment does not change, the Lamarckian concept was that constant development created an escalator-like process, with species constantly moving up the scale (of which human kind currently represented the pinnacle) and being replaced by new, rudimentary life-forms, formed by spontaneous generation in water, joining at the bottom. Lamia does not fit into such a scenario, and if she has to be cast in evolutionary terms she must be deemed a failure, since her mutated state (as a woman) fails to become permanent and is not passed on to any offspring. She is a true monster in the manner of Green's abortive embryos, however, since she has the mixed attributes of different species: not only human and serpentine ones, but also those of other animals: 'Striped like a zebra, freckled like a pard, / Eyed like a peacock' (I. 49–50). Her woman/serpent qualities carry rich allusions to biblical mythology, to classical creatures such as Ovid's Scylla, to Spenser's Duessa, Milton's Sin and Coleridge's Life-in-Death, and to folkloric figures which mix the animal with the human, such as melusines, mermaids, sirens and silkies. She is also close kin to Geraldine in Coleridge's *Christabel*, and in both poems it is the momentary glimpses we catch of the serpent-nature within a human being – the instants when one species seems to show itself incongruously within another – that cause most disturbance:

> A snake's small eye blinks dull and shy;
> And the lady's eyes they shrink in her head,
> Each shrunk up to a serpent's eye,
> And with somewhat of malice, and more of dread,
> At Christabel she looked askance! –
> One moment – and the sight was fled!
> (*Christabel*, 583–8)

> Her head was serpent, but ah, bitter sweet!
> She had a woman's mouth with all its pearls complete.
> (*Lamia*, I. 60)

Lamia's determination to mate out of her kind is deliberately invoked by Keats to disturb and create tension in the reader around sexual topics. Darwin's aim was very much the opposite: to maintain an even, impersonal tone that would conciliate the reader and minimize dismay or disagreement. However, even his determinedly unsensational prose could not avoid raising sexual anxieties in his readers, and Darwin's work, Beer says (*Darwin's Plots,* p. 9),

> foregrounds the concept of kin – and aroused many of the same dreads as fairy-tale in its insistence on the obligations of kinship, and the interdependence between beauty and the beast. Many Victorian rejections of evolutionary ideas register a physical shudder. In its early readers one of the lurking fears it conjured was miscegeny – the frog in the bed – or what Ruskin called 'the filthy heraldries which record the relation of humanity to the ascidian and the crocodile'.[33]

Gothic and some Romantic literature seems to reflect an imaginative environment in which it was possible to play, deliberately and dangerously (but nevertheless light-heartedly) with such ideas: in the way that the Shelleys, Claire Clairmont, Byron and Dr John Polidori purposely excited and frightened themselves with a recitation of *Christabel* and readings of ghost stories at the Villa Diodati in the nights in 1816 that *Frankenstein* was conceived. It *is* possible to read the grotesque and erotic elements of *Lamia* and *Christabel* as beginning to explore a consciousness of the links between humankind and 'the ascidian and the crocodile', but this consciousness seems to be shielded and made safe by the belief, still most widely accepted at the beginning of the nineteenth century, that humankind had been singled out from the rest of external nature (including animals) by a separate act of divine creation and the possession of an immortal soul. As the implications of evolutionary theory became stronger, however (but still many years before the publication of Charles Darwin's work) the consequences of humankind's shared ancestry with animals becomes more threatening. *In Memoriam,* written between 1833 and 1850, can be read as a discussion of the doubts and difficulties involved in accepting human consanguinity with animals, mediated by the possibility of successively higher separate creations and by an exaltation of human spirituality as transcending rational science. This involves an emphatic repudiation of 'the beast':

> Arise and fly
> The reeling Faun, the sensual feast;
> Move upward, working out the beast,
> And let the ape and tiger die,
> (*In Memoriam,* cxvii)

as well as a hope of humankind's elevation to 'the crowning race,' 'No longer half-akin to brute', which is in progress towards

> That God, which ever lives and loves,
> One God, one law, one element,

---

[33] When Godwin tried to persuade Wordsworth to write a versified form of the story of Beauty and the Beast for his Juvenile Library, Wordsworth replied (9 March 1811): 'I confess there is to me something disgusting in the notion of a human being consenting to mate with a beast, however amiable his qualities of heart .'

And one far-off divine event,
To which the whole creation moves.
(In Memoriam, cxxxi)

Another source of anxiety in Tennyson's poetry is the possibility of
humankind being made extinct by the Lamarckian rise of another species: and
it is in this light that the poet reflects upon the fate of the dinosaurs in *Maud*,
iv. 6–11 (published in 1855, still four years before *The Origin):*

A monstrous eft was of old the Lord and Master of Earth,
For him did his high sun flame, and his river billowing ran,
And he felt himself in his force to be Nature's crowning race.
As nine months go to the shaping an infant ripe for his birth,
So many a million ages have gone to the making of man:
He now is first, but is he the last? is he not too base?[34]

In the early part of the century, however, the possible extinction of certain
human 'species' can still be treated as a joke, as it is by Keats in his remarks
quoted above about parsons and lawyers, and by Byron, when he envisages
how the future 'new worldlings of the then new East' will dig up the bones of
the whale-like George IV, debate 'how such animals could sup' and discover
that

these great relics when they see 'em
Look like the monsters of a new Museum.
(Don Juan, IX. 291–320)

In contrast to this comic mode, Keats's 'Ode to a Nightingale' envisions no
evolving perfection for the poet, but a painful transience in which, to the bird's
'high requiem', he will have 'become a sod'; while the 'hungry generations' of
the poem seem to recall Malthus's arguments about the difficulty of sustaining
human populations from the earth's finite resources. The Ode by-passes the
traditional iconography which makes birds and butterflies symbols of the
immortality of the soul, focusing instead on the death of the poet, and on a non-
religious 'immortality' for the bird which is comforting precisely because it
implies stasis, rather than development:

Thou wast not born for death, immortal Bird!
No hungry generations tread thee down;
The voice I hear this passing night was heard
In ancient days by emperor and clown:
Perhaps the self-same song that found a path
Through the sad heart of Ruth ...
('Ode to a Nightingale', 61–6)

---

[34] Christopher Ricks's commentary in his edition of *The Poems of Tennyson*, II: 530,
demonstrates the influence of Robert Chambers's *Vestiges of the Natural History of Creation*
(1844) on this passage: 'Are there yet to be species superior to us in organization, purer in
feeling, more powerful in device and act, and who shall take a rule over us?'

The 'Ode on a Grecian Urn' celebrates the stasis that is also inherent to a work of art, choosing to praise an ancient artefact (although a modern one might be expected to be more 'perfect' according to the evolutionary theories Keats had been taught) and again contrasting it with the transience and mortality of the poet himself:

> Thou, silent form, dost tease us out of thought
> As doth eternity. Cold pastoral!
> When old age shall this generation waste,
>   Thou shalt remain, in midst of other woe
> Than ours, a friend to man.
> ('Ode on a Grecian Urn', 44–8)

Both poems look for their sources of permanence outside the (Christian) religious sphere. Neither of them expresses any expectation of human perfectibility, nor any hope that evolution – or any other natural process – will lead to higher forms and species, or to a physical or spiritual transcendence.

Degeneration

Byron's work, too, expresses a fully-fledged pessimism about human perfectibility, of which the human 'degeneration' motif in his 1821 'mystery' play *Cain* is the clearest manifestation. Byron claimed in the Preface to the play that the ideas and fictions of Lucifer were not a reflection of his own beliefs, and that he himself was 'no enemy to religion but the contrary' (letter to Thomas Moore, *BLJ*, IX. 118). Nevertheless, the dramatic form provides Byron in all his late 'metaphysical' plays *(Cain, Heaven and Earth* and *The Deformed Transformed)* with a useful vehicle in which to explore and give imaginative life to some highly contentious ideas which, by drawing attention to particular elements of the evolutionary debate of his time, were widely considered to undermine religion. Contemporary critics described the play's impiety as 'frightful', calling it 'a heinous offence against society' and 'Hideous Blasphemy'; and even Byron's friend and fellow poet Thomas Moore described the Catastrophism on which *Cain* is based, in a letter to Byron, as 'most desolating ... in the conclusions to which it may lead some minds,' and regretted the fact that Byron had 'carr[ied] this deadly chill' into the minds of 'the young, the simple'.[35]

Byron adopted his ideas about Catastrophism from Cuvier's *Recherches sur les ossemens fossiles de quadrupèdes* (1812). Cuvier's work posits, as Byron put it in a prefatory note to the play, 'that the world had been destroyed several times before the creation of man,' leaving in the wake of these upheavals 'different strata and bones of enormous and unknown animals found in them'. This speculation does not in itself challenge Christian orthodoxy, since it is, as Byron points out, 'not contrary to the Mosaic account, but rather confirms it; as no human bones have yet been discovered in those strata, although those of

---

[35] *Byron, The Critical Heritage* ed by Andrew Rutherford (London: Routledge & Kegan Paul, 1970) pp. 217, 219. *The Letters of Thomas Moore,* ed by Wilfred S. Dowden, 2 vols (Oxford: Oxford University Press, 1964) II. 505.

many known animals are found near the remains of the unknown.' Indeed, although not apparently intended so by Cuvier (who was purportedly a religious sceptic or even an atheist), Catastrophism was presented to English-speaking readers through Robert Kerr's translation (*Essay of the Theory of the Earth*, 1813), and through the Preface and notes by Robert Jameson, as a theologically highly conservative system.[36] Subjects so important could not fail, Jameson claimed (Preface, p. ix), 'to admonish the sceptic, and afford the highest pleasure to those who delight in illustrating the truth of the Sacred Writings, by an appeal to the facts and reasonings of natural history.'

The following year, however, the Scottish theologian Dr Thomas Chalmers (later the author of the first Bridgewater Treatise) devoted a chapter of his *Evidence and Authority of the Christian Revelation* to 'the scepticism of geologists' and argued that, judging only by his attributes as known from the natural world, God could be devoid of moral characteristics, and that the revelation of scripture was required to prove the nature of his goodness.[37] Then in 1819 a refutation of Chalmers and a renewed attempt at an orthodox interpretation of Cuvier was made by the Revd William Buckland (later Dean of Westminster) in his inaugural lecture as the first Reader in Geology at the University of Oxford: a post endowed by the Prince Regent. Buckland dismissed any 'difficulty from an apparent nonconformity of certain Geological phenomena with the literal and popular account of the creation, as it is presented to us in the book of Genesis' and stated his objective as being 'to show that the study of geology has a tendency to confirm the evidence of natural religion, and that the facts developed in it are consistent with the accounts of the creation and deluge recorded in the Mosaic writings'.[38]

Cain can thus be seen as Byron's response to this debate: his commentary upon the various interpretations of the geological phenomena, and on the patterns of resemblance imposed upon the 'scientific' discoveries by theological conservatives.[39] Byron may have gleaned his scepticism about the potential for over-enthusiastic geological analogies, and the tendency to leap to conclusions on the basis of only minimal evidence, from Cuvier himself, who described Lamarck and other scientists (p. 45) as 'carried away by ... bold or extravagant conceptions'; and complained (pp. 48–9) that 'the conditions of the problem never having been taken into consideration ... it has remained hitherto

---

[36] For Cuvier's atheism, see Martin J. Rudwick, *Georges Cuvier: Fossil Bones and Geological Catastrophes: New Translations and Interpretations of the Primary Texts* (Chicago: University of Chicago Press, 1997) p. 259.

[37] *The Evidence and Authority of the Christian Revelation* (Edinburgh, Blackwood, 1814) p. 223: 'We profess ourselves too little acquainted with the character of God; and that in this little corner of his works, we see not far enough to offer any decision on the merits of a government, which embraces worlds, and reaches eternity.' In 1843 Chalmers led the Disruption which caused the formation of the Free Church in Scotland.

[38] William Buckland, *Vindiciae Geologicae, or, The Connexion of Geology with Religion Explained* (Oxford: Oxford University Press, 1820) p. 22 and Dedication.

[39] As Buckland mentions in the appendix to his lecture, Chalmers's *Evidence and Authority* was reviewed in the *Quarterly Review* XVII. 34 (July 1817) pp. 451–63. The reviewer drew attention to Chalmers' unflattering picture of the Deity. Byron thanked John Murray for sending him copies of 'the Edin[burg]h and Quarterly' on 4 September 1817 (*BLJ*, V. 262).

indeterminate, and susceptible of many solutions – all equally good, when such or such conditions are abstracted; and all equally bad, when a new condition comes to be known'.[40] Neither a geologist nor a theologian himself, and working (as he said) as 'an imaginative man', Byron creates in *Cain* an elaborate alternative fiction which phantasizes ebulliently on the known 'facts' (both those presented by the scientists and those in the biblical account).[41] In doing so, it ironically critiques not only the specific controversy about the complementarity between Cuvier's theories and the Mosaic writings, but also the way in which both the geologists and theologians were creating huge 'systems' of scientific theory or belief in this area out of the flimsiest of material.

Natural theology

Buckland claimed in his 1819 lecture that, since it demonstrated how the geological account supported the 'truth' of the Mosaic text on the deluge, 'Geology contributes proofs to Natural Theology strictly in harmony with those derived from other branches of natural history' (p. 18). As read by Jameson and Buckland, Catastrophism did indeed support the prevailing concepts of Paley's thesis that the existence of design in nature was unanswerable proof of the existence of a deity with specifically beneficent intentions towards mankind. Byron recorded 'Paley' in a list of books he had read by November 1807, and it is highly likely that he was introduced to Paley's *Natural Theology* as an undergraduate at Cambridge, since it was his tutor at Trinity, the Revd Thomas Jones, who had been responsible in the 1780s for introducing Paley's other works into the University syllabus (where they remained until 1920).[42] Darwin, similarly, encountered Paley at Cambridge:

---

[40] Cuvier also cast doubt on the reliability of the biblical account by referring (p. 152) to 'the incoherence of all these traditionary tales' which 'attest the barbarism and ignorance of all the tribes around the Mediterranean'.

[41] Letter to Thomas Moore about *Cain,* 4 March, 1822 *(BLJ,* IX. 118–9): 'like all imaginative men, I, of course, embody myself with the character while I draw it, but not a moment after the pen is from off the paper.'

[42] See *BCMP* p. 5; M. L. Clarke, *Paley: Evidences for the Man* (London: SPCK, 1974) p. 127, and Marchand, *Byron a Biography,* I. 102–3. Paley's *Natural Theology* was never part of the syllabus at Cambridge, although his other works were. See Aileen Fyfe, 'The Reception of William Paley's *Natural Theology* in the University of Cambridge', *British Journal of the History of Science,* XXX (1997) pp. 321–35 (p. 321): 'Many scholars have assumed that it *[Natural Theology]* was a set text at the university in the early nineteenth century. However, a study of the examination papers of the university, and contemporary memoirs, autobiographies and correspondence, reveals no evidence that this was so, though it did appear in some of the college examinations.' As a nobleman, Byron's final examination at Cambridge was limited to a few minutes' 'disputation' with his tutor and the paying of a fee, since Thomas Jones's attempt to introduce full examinations for noblemen had been turned down by the University in 1787. See Anne Barton, 'Lord Byron and Trinity', p. 4.

In order to pass the B.A. examination, it was, also, necessary to get up Paley's *Evidences of Christianity,* and his *Moral Philosophy.* This was done in a thorough manner, and I am convinced that I could have written out the whole of the *Evidences* with perfect correctness, but of course not in the clear language of Paley. The logic of this book and as I may add of his *Natural Theology* gave me as much delight as did Euclid. The careful study of these works, without attempting to learn any part by rote, was the only part of the academical course which, as I then felt and as I still believe, was of the least use to me in the education of my mind. I did not at that time trouble myself about Paley's premises; and taking these on trust I was charmed and convinced by the long line of argumentation.[43]

Byron's response to Paley, as manifested in his 'metaphysical' plays, is essentially the same as that later fully articulated by Chalmers: that the God demonstrated by nature is not necessarily a beneficent designer, and may indeed be a malicious and vindictive creator and destroyer:

> The proof both for intelligence and power may be as complete with one set of ends as with another set wholly opposite. There may be as thorough an impress of skill and energy on a machinery of torture, as on some bland and beneficent contrivance that operates a blessing throughout the sphere of its activity – on the structure, for example, of a serpent's envenomed tooth, as on the structure of those teeth which prepare the aliment for digestion, and subserve one of the most useful functions of the animal economy. It is thus that a wicked and malignant spirit could give decisive, but most terrible demonstration withal of his Natural Attributes – so that these on the one hand may be most strikingly and satisfactorily evinced, while the Moral Attributes of the other may be involved in the Mystery of those contradictory appearances of nature, which the wisdom of man has so vainly endeavoured to unravel.[44]

Byron pushed the challenge to more extreme and unorthodox conclusions than Chalmers, and his literalistic treatments of the Bible (which expose its apparent limitations and contradictions as a factual record) challenged revealed, as well as natural, religion. In the same way as for Darwin in later decades, however, Natural Theology forms a basis for Byron's work in his

---

[43] *The Autobiography of Charles Darwin* ed by Nora Barlow (London: Pickering, 1989) p. 101.

[44] 'On Natural Theology,' vols I and II of *The Works of Thomas Chalmers,* 25 vols (Glasgow: William Collins, 1836–42) I. 365. Byron in 1823 described the Scottish minister as 'an infidel, as every man had the right to be' (Lovell, *His Very Self,* p. 421). Robert Browning's 'Caliban Upon Setebos; or, Natural Theology in the Island' (published in 1864) provides another instance of imaginative argument with Paley (influenced by the 'Bridgewater Treatises', including that of Chalmers – and also probably by Darwin, whom Browning is thought to have read in the early 1860s). Caliban's musings on the relationship between God, Man and the natural world debate the supposed beneficence or malignity of God and ironize the theologians' creation of the deity in their own image. Browning owned Paley's *A View of the Evidences of Christianity:* see *The Complete Works of Robert Browning* ed by John C. Berkey, Allan C. Dooley and Susan C. Dooley, 9 vols (Athens, Ohio: Ohio University Press, 1996) VI. 446).

metaphysical plays in the sense that he is constantly imaginatively debating or arguing with it. Darwin wrote in his *Autobiography* (ed by Barlow, p. 120):

> The old argument of design in nature, as given by Paley, which formerly seemed to me so conclusive, fails, now that the law of natural selection has been discovered. We can no longer argue that, for instance, the beautiful hinge of a bivalve shell must have been made by an intelligent being, like the hinge of a door by man.

It can be said of Byron as it has been of Darwin, that he took Paley's answers and converted them into questions.[45]

Interrogating natural theology: man de-centralized

There are three particular areas in which I want to look at Byron's interrogation of Natural Theology and to relate it to Darwinism: first, the displacement of humankind from a central position in the universe, and its corollary a raising of the status of other species and a promotion of a sense of kinship between humans and animals; second, the problem of pain and suffering in all species, in relation to a supposedly beneficent creator; and third, the issues of design, perfection and deformity. All three of these areas provide a commentary on early-nineteenth-century perceptions of the relations between humankind and other animals.

In the prefatory note to *Cain* Byron claimed that 'The assertion of Lucifer, that the pre-Adamite world was also peopled by rational beings much more intelligent than man, and proportionately powerful to the mammoth,' is 'of course, a poetical fiction to help him make out his case'. Diabolical fiction or not, however, this proposition forms part of Byron's deliberate de-centring of man in the universe, and what Beer *(Darwin's Plots,* p. 21) says of Darwin is equally true of Byron in his metaphysical plays: that they 'demonstrate in their major narratives of geological and natural history that it was possible to have plot without man – both plot previous to man and plot even now regardless of him'. Cain describes the pre-Adamite beings as

> Haughty, and high, and beautiful, and full
> Of seeming strength, but of inexplicable
> Shape; for I never saw such. They bear not
> The wing of seraph, nor the face of man,
> Nor form of mightiest brute, nor aught that is
> Now breathing; mighty yet and beautiful
> As the most beautiful and mighty which
> Live, and yet so unlike them, that I scarce
> Can call them living.
> *(Cain,* II. ii. 54– 62)

These superhuman and intelligent creatures were also imagined by Shelley, whose Panthea in *Prometheus Unbound* (IV. 296–303) speaks of:

---

[45] Robert M. Young, *Darwin's Metaphor, Nature's Place in Victorian Culture* (Cambridge: Cambridge University Press, 1985) p. 39.

> The wrecks beside of many a city vast
> Whose population which the earth grew over
> Was mortal, not but human; see, they lie
> Their monstrous works, and uncouth skeletons
> Their statues, homes and fanes; prodigious shapes
> Huddled in grey annihilation, split,
> Jammed in the hard, black deep.

Both Byron and Shelley may have been influenced by William Beckford who, in *Vathek,* has the Satanic Eblis show Vathek and Nouronihar the treasures of pre-Adamite sultans, together with 'the various animals that inhabited the earth prior to the creation of that contemptible being whom ye denominate the father of mankind'.[46] Lucifer claims that these 'Living, high, / Intelligent, good, great, and glorious things,' *(Cain,* II. i. 67–8) were not only 'As much superior unto all thy sire, / Adam, could e'er have been in Eden,' (69–70) but also as much superior to Cain as he will be to his own descendants: 'The sixty-thousandth generation ... / In its dull damp degeneracy' (71–2). *Don Juan* recounts the same story in a comic mode, foreseeing a future time

> When this world shall be *former,* underground,
>   Thrown topsy-turvy, twisted, crisped and curled,
> Baked, fried, or burnt, turned inside-out, or drowned,
>   Like all the worlds before, which have been hurled
> First out of then and back again to Chaos,
> The Superstratum which will overlay us.
>
> So Cuvier says; – and then shall come again
>   Unto the new Creation, rising out
> From our old crash, some mystic, ancient strain
>   Of things destroyed and left in airy doubt:
> Like to the notions *we* now entertain
>   Of Titans, Giants, fellows of about
> Some hundred feet in height, *not* to say *miles,*
> And Mammoths, and your winged Crocodiles.

'Even worlds miscarry, when too oft they pup,' the narrator claims,

> And every new Creation hath decreased
> In size, from overworking the material –
> Men are but maggots of some huge Earth's burial.
> *(Don Juan,* IX. 291–312)

The principle of 'degeneration' which Lucifer and Byron apply so devastatingly to humankind is based on a term widely used by French naturalists. As employed by Buffon, Cuvier and Lamarck, and later by Darwin, it refers to the way in which a 'genus' of animals is changed by evolution: a development or moving away from the original type. Thus Clarence J. Glacken describes how:

---

[46] See McGann, *BCPW,* VI. 655.

Buffon attached great importance to the 'degeneration' of animals; by this term he meant the variability of a species under the influence of climate and food; in the case of the domestic animals, there was the added influence of the 'yoke of slavery'. Degeneration was even more marked among the domestic animals reduced to slavery (a favorite word of Buffon's).[47]

In line with his theriophilic predisposition, Byron deploys the French word in a literal English sense to suit his (Rousseauian and satirical) purpose of demonstrating that modern, socialized, human beings are physically and morally 'degenerate' creatures, especially compared with wild animals.[48] This belittling of humankind is a key Byronic motif, especially in the late plays where ironic and demonic commentators such as Lucifer and the Stranger/Caesar are able to anatomize man from a point of view outside humankind and with a vantage of infinity:

> CAESAR   Well! I must play with these poor puppets: 'tis
> The spirit's pastime in his idler hours.
> When I grow weary of it, I have business
> Among the stars, which these poor creatures deem
> Were made for them to look at. 'Twere a jest now
> To bring one down amongst them, and set fire
> Unto their ant hill.
> *(The Deformed Transformed,* I. i. 320–6)

Byron offsets this diabolical mockery by stressing the poignant tenderness of humanity:

> ADAH   Look! how he laughs and stretches out his arms,
> And opens wide his blue eyes upon thine,
> To hail his father, while his little form
> Flutters as wing'd with joy.
> *(Cain,* III. i. 149–52)

Even the beauty of the universe, however, emphasizes the point that humankind is only a tiny and insignificant part of it:

> CAIN   Oh, thou beautiful
> And unimaginable ether! and

---

[47] Clarence J. Glacken, *Traces on the Rhodian Shore: Nature and Culture in Western Thought from Ancient Times to the End of the Eighteenth Century* (Berkeley: University of California Press, 1967) p. 674. See Buffon's 'De la dégénération des animaux,' *Histoire naturelle,* XIV. 317.

[48] See Rousseau, *Discourse on the Origins and Foundations of Inequality among Men* (Harmondsworth: Penguin, 1984) p. 52: 'The horse, the bull, and even the ass are generally of greater stature, and always more robust, and have more vigour, strength and courage, when they run wild in the forests than when bred in the stall. ... To this it may be added that there is still a greater difference between savage and civilized man than between wild and tame beasts: for men and brutes having been treated alike by nature, the several conveniences in which men indulge themselves still more than they do their beasts, are so many additional causes of their deeper degeneracy.'

Ye multiplying masses of increased
And still-increasing lights! What are ye? what
Is this blue wilderness of interminable
Air, where ye roll along? ...

LUCIFER   [L]ook back to thine earth!

CAIN   Where is it? I see nothing but a mass
Of most innumerable lights. ...
Why, I have seen the fire-flies and fire-worms
Sprinkle the dusky groves and green banks
In the dim twilight, brighter than yon world
Which bears them.

LUCIFER   Thou hast seen worms and worlds . . .
          (Cain, II. i. 98–126)

It was the problem of the countless stars disclosed by the telescope, and
their complete irrelevance from a human point of view, which had compelled
Robert Boyle in the seventeenth century to abandon the notion that everything
in the universe was created for man's benefit alone. Paley admitted that
astronomy 'is not the best medium through which to prove the agency of an
intelligent Creator' *(Natural Theology,* p. 409), while Tom Paine stated baldly
in *The Age of Reason* (p. 277) that 'to believe that God created a plurality of
worlds at least as numerous as what we call stars, renders the Christian system
of faith at once little and ridiculous and scatters it in the mind like feathers in
the air.' Byron commented in a letter of 18 June 1813 to William Gifford: 'It
was the comparative insignificance of ourselves & our world when placed in
competition with the mighty whole of which it is an atom that first led me to
imagine that our pretensions to eternity might be overrated' *(BLJ,* III. 64).

Such musings on the stars are echoed by Darwin in an early notebook:
'Mayo (Philosophy of Living) quotes Whewell as profound because he says
length of days adapted to duration of sleep in man!! whole universe so
adapted!! and not man to Planets – instance of arrogance!!'[49] Beer points out
that

> Man is a determining absence in the argument of *The Origin of Species*
> ... He appears only once in the first edition as the subject of direct
> inquiry. ... the exclusion had an immediate polemical effect: it removed
> man from the centre of attention.[50]

This 'removing of man from the centre of attention' has the effect in both
Darwin and Byron of an accompanying emphasis on other species. Darwin,
for instance, notes *(Early Notebooks,* p. 446): 'People often talk of the
wonderful event of intellectual man appearing. – the appearance of insects with
other senses is more wonderful. its mind more different probably &

---

[49] *Darwin's Early and Unpublished Notebooks,* ed by Paul H. Barrett (London: Wildwood House, 1974) p. 455.
[50] Gillian Beer, 'The Face of Nature': Anthropomorphic Elements in the Language of *The Origin of Species',* in *Languages of Nature: Critical Essays on Science and Literature,* ed by Ludmilla Jordanova (London: Free Association Books, 1986) pp. 207–43 (pp. 212–3).

introduction of man nothing compared to the first thinking being.' In *The Island,* Byron denies the force of the pathetic fallacy at a key moment in the drama by turning his attention away from the smashed bodies of Christian Fletcher and his comrades on the rocks to the quite unrelated concerns of the natural world:

> Cold lay they where they fell, and weltering,
> While o'er them flapp'd the sea-bird's dewy wing,
> Now wheeling nearer from the neighbouring surge,
> And screaming high their harsh and hungry dirge:
> But calm and careless heaved the wave below,
> Eternal with unsympathetic flow;
> Far o'er its face the dolphins sported on,
> And sprung the flying fish against the sun,
> Till its dried wing relapsed from its brief height,
> To gather moisture for another flight.
> *(The Island,* IV. xiii. 363–72)

The emphasis may also be on the superior abilities of other species: Lucifer (and Byron, in the preface to *Cain)* insist that the tempter of Eden was a real serpent, not an evil spirit, more advanced than human beings in 'wisdom' or intelligence:

> The snake was the snake –
> No more; and yet no less than those he tempted,
> In nature being earth also – *more* in *wisdom,*
> Since he could overcome them, and foreknew
> The knowledge fatal to their narrow joys.
> *(Cain,* I. i. 223–7)

While in Tennyson there is a need to repudiate the beast in order to emphasize the special status of humankind, in Byron and Darwin the concept of kinship between humans and animals is not necessarily one which is disadvantageous to man, since animals may be in some ways superior.

The problem of pain

Lucifer's predominant aim for Cain is, as Byron explained to John Murray,

> to depress him still further in his own estimation than he was before –
> by showing him infinite things – & his own abasement – till he falls into
> the frame of mind – that leads to the Catastrophe ... from rage and fury
> against the inadequacy of his state to his conceptions. *(BLJ,* IX. 3–4)

Lucifer does *not* do this (as might be expected) by convincing Cain that he is bestial – animal-like – in his attributes or origins: instead, there is a remarkable degree of concurrence between the man and the devil in the play to the effect that humans and animals are united in being the innocent victims of a cruel God, and that their common suffering is expressed by their fellowship. Darwin

agrees: as he puts it, animals and humans are 'melted together' by pain and
suffering:

> If we choose to let conjecture run wild, then animals, our fellow
> brethren in pain, disease, death, suffering and famine – our slaves in the
> most laborious works, our companions in our amusements – they may
> partake [of?] our origin in one common ancestor – we may be melted
> together.[51]

Cain is driven to murder partly by Jehovah's approval of Abel's animal
sacrifice:

> CAIN   *His pleasure!* what was his high pleasure in
>   The fumes of scorching flesh and smoking blood,
>   To the pain of bleating mothers, which
>   Still yearn for their dead offspring? or the pangs
>   Of the sad ignorant victims underneath
>   Thy pious knife? Give way! This bloody record
>   Shall not stand in the sun, to shame creation!
>   *(Cain,* III. i. 298–305)

One of the most insistent themes of the work (that of the injustice of the
innocent suffering on behalf of the guilty) is articulated in terms of animals and
sacrifice; and the 'fumes of scorching flesh and smoking blood' on Abel's altar
become a symbol for the other kinds of 'Molochism' that the play deplores:
the suffering of children on behalf of their parents, of descendants for their
ancestors, and (most notoriously) of Christ on the cross for the 'sins of the
whole world':

> LUCIFER   perhaps he'll make
>   One day a Son unto himself – as he
>   Gave you a father – and if he so doth
>   Mark me! – that Son will be a Sacrifice.
>   *(Cain,* I. i. 163–6)[52]

Cain perceives animals as suffering unjustly as a result of human disobedience
in eating the fruit of the tree of knowledge:

> CAIN   But animals –
>   Did they too eat of it that they must die?
>
> LUCIFER   Your Maker told ye, *they* were made for you,
>   As you for him. – You would not have their doom

---

[51] 'Notebook 1', 1837–38, from *The Life and Letters of Charles Darwin; Including an
Autobiographical Chapter,* ed by Francis Darwin, 3 vols (London: John Murray, 1888) II. 6.
Other editors have read this as 'netted together'.

[52] Moloch, god of the Ammonites, to whom children were sacrificed according to II *Kings*
22. 10, receives five mentions in Byron's poetry, usually as a short-hand reference to generals
and political leaders who are prepared to sacrifice their troops or their people in war. Tom
Paine (p. 256) also deplored this element of Christianity: 'Moral justice cannot take the
innocent for the guilty, even if the innocent would offer itself'.

Superior to your own? Had Adam not
Fallen, all had stood.

CAIN   Alas! the hopeless wretches!
           (Cain, II. ii. 152–7)

This theme is more fully explored through Cain's description of Adam healing a lamb, which has been stung by a reptile:

CAIN   the poor suckling
           Lay foaming on the earth, beneath the vain
           And piteous bleating of its restless dam;
           My father pluck'd some herbs and laid them to
           The wound; and by degrees the helpless wretch
           Resumed its careless life, and rose to drain
           The mother's milk, who o'er it tremulous
           Stood licking its reviving limbs with joy.
           Behold, my son! said Adam, how from evil
           Springs good!

But, Cain comments:

           I thought, that 'twere
           A better portion for the animal
           Never to have been stung at all, than to
           Purchase renewal of its little life
           With agonies unutterable, though
           Dispell'd by antidotes.
           (Cain, II. ii. 290–305)

Jerome McGann suggests that one of Byron's sources for this incident is the doctrine of the *felix culpa* ('the fortunate fault') in the article on 'The Paulicians' in Pierre Bayle's *Dictionary.* This doctrine, Bayle says, is like 'comparing God ... with the father of a family, who should suffer his children's legs to be broke, in order to shew what an able bone-setter he is'.[53]

The existence of pain and suffering was perceived as one of the greatest challenges to Natural Theology but also as an opportunity for its greatest triumph. Paley, for instance, used it as a means of instancing and substantiating the goodness of God:

The annexing of pain to the means of destruction is a salutary provision, inasmuch as it teaches vigilance and caution: both gives notice of danger, and excites those endeavors which may be necessary to preservation. ... Pain also itself is not without its *alleviations.* It may be violent and frequent, but it is seldom both violent and long continued; and its pauses and intermissions become positive pleasures. It has the power of shedding a satisfaction over intervals of ease, which I believe

---

[53] From the 1734 English translation of *Mr Bayle's Historical and Critical Dictionary,* quoted in *BCPW,* VI. 661.

few enjoyments exceed. ... I am far from being sure that a man is not a gainer by suffering a moderate interruption of bodily ease for a couple of hours out of the four and twenty. *(Natural Theology,* pp. 531–3)

Byron acknowledged to Teresa Guiccioli in 1822 that it could 'hardly be conceived how much [he] suffer[ed] at times' from the pain of his deformed foot, and a doctrine which taught that bodily pain was in any way of benefit for the betterment of humanity evidently seemed to him utterly repugnant.[54] Like Byron, Darwin thought that the existence of animal suffering disproved Natural Theology's doctrine that pain was useful as a teacher, and he could also go further than Byron in explaining – if not the metaphysical 'reason' for suffering – at least the way in which it operated through the struggle for existence:

That there is much suffering in the world no one disputes. Some have attempted to explain this with reference to man by imagining that it serves for his moral improvement. But the number of men in the world is as nothing compared with that of all the other sentient beings, and they often suffer greatly without any moral improvement. This very old argument from the existence of suffering against the existence of an intelligent First Cause seems to me a strong one; whereas, as just remarked, the presence of much suffering agrees well with the view that all organic beings have been developed through variation and natural selection. (Darwin, 'Autobiography', from *Life and Letters,* 1888, I. 311)

The Darwinian world demonstrated in even more acute form than the Byronic what Thomas Huxley called 'the moral indifference of nature' and 'the unfathomable injustice of the nature of things' (cited by Brooke, p. 92). Although in public Darwin avoided controversy over this issue, in letters such as one to the sympathetic American naturalist, Asa Gray, he was prepared to own that

I cannot see, as plainly as others do, & as I shd wish to do, evidence of design & beneficence on all sides of us. There seems to me too much misery in the world. I cannot persuade myself that a beneficent & omnipotent God would have designedly created the Ichneumonidae with the express intention of their feeding within the living bodies of caterpillars, or that a cat should play with mice. Not believing this I see no necessity in the belief that the eye was expressly designed.[55]

Darwinism provides its own reasons for death and suffering, and magnifies the incidence of struggle because it is an essential part of Darwin's version of the mechanism of the origin of species. The Darwinian reasons for death and

---

[54] Marchand *Biography* III. 1047. Byron wrote in his 'Detached Thoughts' of 1821–22 (no. 96) that 'A material resurrection [of the body] seems strange and even absurd except for purposes of punishment' *(BLJ,* IX. 45).

[55] *The Correspondence of Charles Darwin,* ed by Frederick Burkhardt and others, 8 vols (Cambridge: Cambridge University Press, 1985–93) VIII. 224. This echoes Erasmus Darwin in *The Temple of Nature,* IV. 33–4: 'The wing'd Ichneumon for her embryon young / Gores with sharp horn the caterpillar throng.'

suffering are, of course, non-religious and non-human-centred, but nevertheless he often chose to express them through images of warfare:

> De Candolle, in an eloquent passage, has declared that all nature is at war, one organism with another, or with external nature. Seeing the contented face of nature, this may at first well be doubted; but reflection will inevitably prove it to be true.[56]

> What a struggle between the several kinds of trees must have gone on during long centuries, each annually scattering its seeds by the thousand; what war between insect and insect – between insects, snails and other animals with birds and beasts of prey – all striving to increase, and all feeding on each other or on the trees or their seeds and seedlings, or on the other plants which first clothed the ground and thus checked the growth of the trees! *(The Origin,* p. 126)

Byron's preoccupation with war extends from the early cantos of *Childe Harold* through to his last, unfinished, play *The Deformed Transformed,* and his works portray struggle and warfare as at its fiercest when it is complicated by kinship. Similarly, Darwin noted the way in which the fiercest competition in nature is actually closest to home: 'the struggle almost invariably will be the most severe between the individuals of the same species, for they frequent the same districts, require the same food, and are exposed to the same dangers' *(The Origin,* p. 128). In *Cain* the individual and personal struggle within the archetypal family, and particularly that between brothers, is presented as the prototype of human wars to come, and Byron called the familial situation of the play 'the Politics of Paradise' *(BLJ,* VIII. 216). In *Heaven and Earth,* which picks up the biblical story a few generations later, just before the Flood, the issue of kinship and struggle is set in the wider context of a world inhabited by several different humanlike groups, which are related but in the process of becoming alienated and separated as if into different 'species'. Only one of these groups (the descendants of Noah) has been selected on the basis of its genes (rather than its virtue) to be saved from the forthcoming catastrophe. The demonic race of spirits – victims of a previous Jehovan purge – proclaim as the catastrophe gathers the strength of their own kinship:

> SPIRIT   Who would outlive their kind,
> Except the base and blind?
> Mine
> Hateth thine
> As of a different order in the sphere,
> But not our own.
> There is not one who hath not left a throne
> Vacant in heaven to dwell in darkness here,
> Rather than see his mates endure alone.
> *(Heaven and Earth,* III. 145–53)

[56] *Foundations of the Origin of Species,* ed by Francis Darwin (London: Pickering, 1986) p. 68.

The coming time, when it seems the human race will become extinct, is celebrated by the spirits in terms which echo the palaeontological sources Byron used:

SPIRIT The creatures proud of their poor clay,
        Shall perish, and their bleached bones shall lurk
        In caves, in dens, in clefts of mountains, where
        The Deep shall follow to their latest lair.
        (Heaven and Earth, III. 173–6)

There are already deep divisions between the two strictly human races (the descendants of the two surviving sons of Adam, Seth and Cain). Additionally, some of the Cainite women (the 'Daughters of Men') have mated with angels (the 'Sons of God') to produce yet another group: 'the glorious giants, who / Yet walk the world in pride' (III. 131–2); and the play's action shows us, in addition, two different 'races' of angels. The complexity of the distinctions between these humanlike groups is contrasted with a situation where the supposedly greater distinctions between men and animals, and between different types of animals, are obliterated, as 'The dragon crawls from out his den, / To herd in terror innocent with men' (III. 798–9). The Isaiahan prophecy of the coming of the Messiah is parodied in words which reflect palaeontological accounts of the way the bones of different species were found jumbled together in caves:

        even the brutes in their despair,
    Shall cease to prey on man and on each other,
    And the striped tiger shall lie down to die
    Beside the lamb, as though he were his brother.
    (Heaven and Earth, III. 177–81)

The play follows the biblical account of the foundation of animal species by their pairing in the Ark, but subjects *Homo Sapiens* to a Linnaean or Darwin-like exploration of the taxonomy of kinship, showing how the fate of its humanlike protagonists is defined not by their individual feelings or qualities, which are often at odds with their descent, but by their race. The pious characters of the play state that matings or miscegenations between the different rational 'species' are a sign of rebellion against God:

JAPHET   for unions like to these
        Between a mortal and an immortal, cannot
        Be happy or be hallow'd,
        (Heaven and Earth, III. 369–71)

and Noah wishes to treat human beings like the animals he couples in the Ark:

        Woe, woe, woe to such communion!
    Has not God made a barrier between earth
    And heaven, and limited each, kind to kind?
    (Heaven and Earth, III. 474–6)

But, as in *Cain,* where Cain and Adah's union is presented as incestuous, but genealogically necessary (and happy) the playwright's sympathies evidently endorse the deviance of those who choose to make love, rather than war, across the bounds of race and even of 'species'.

Byron's last, unfinished, play, *The Deformed Transformed,* mirrors and reverses the situation in *Cain* by providing a setting which is multi-national (the sack of Rome by French and German troops in 1527) but where the politics are presented by the demonic commentator Caesar as being personal and familial. He reminds us that Rome owes its foundation to one brother's murder of another, and that the city has continued to be the site of fratricidal struggle:

> CAESAR   I saw your Romulus (simple as I am)
> Slay his own twin, quick-born of the same womb,
> Because he leapt a ditch ('twas then no wall,
> Whate'er it now be;) and Rome's earliest cement
> Was brother's blood; and if its native blood
> Be spilt till the choked Tiber be as red
> As e'er 'twas yellow, it will never wear
> The deep hue of the Ocean and the Earth,
> Which the great robber sons of Fratricide
> Have made their never-ceasing scene of slaughter
> For ages.
> *(Deformed Transformed,* I. ii. 80–91)

A common influence on Byron's and Darwin's perception of internecine struggle is Malthus's 'Essay on the Principle of Population', which emphasizes the inglorious and economic reasons for warfare. It is on these that Byron sardonically reflects in a Malthusian reference in *Don Juan* Canto I (1043–6):

> The population there [in America] so spreads, they say,
> 'Tis grown high time to thin it in its turn,
> With war, or plague, or famine, any way,
> So that civilisation they may learn;

and these which Darwin applied to the observations he had made during the *Beagle* voyage:

> In October 1838, that is, fifteen months after I had begun my systematic enquiry, I happened to read for amusement Malthus on *Population,* and being very well prepared to appreciate the struggle for existence which everywhere goes on from long-continued observation of the habits of animals and plants, it at once struck me that under these circumstances favourable variations would tend to be preserved, and unfavourable ones to be destroyed. The result of this would be the formation of new species. Here, then, I had at last got a theory by which to work. *(Autobiography,* ed by Barlow, p. 144)

Design and deformity

Within Natural Theology, the existence of war could be explained as a divine necessity (Malthus was a clergyman), but other natural features of the universe, such as earthquakes and painful disease, were embarrassing to the natural theologians. Also acutely embarrassing to Natural Theology were monstrous births and deformities, since they appeared to demonstrate nature in error ('errores naturae') or in sport ('lusus naturae'). Paley's most famous arguments were grounded specifically in the physical perfection of human and animal anatomy (particularly its bilateral symmetry, and the structure of the eye, the hand and the ankle) since he believed that such fitness to purpose could not have been developed as a result of chance, and thus provided clear evidence of the existence of a divine designer. The muscular arrangement of the ankle was, for example, 'so decidedly a mark of intention, that it always appeared to me to supersede, in some measure, the necessity of seeking for any other observation upon the subject' *(Natural Theology,* p. 155).

Paley further believed that his example precisely contradicted the proto-Lamarckian opinion, held by Erasmus Darwin, 'that the parts of animals may have been all formed by what is called *appetancy,* i.e. endeavour, perpetuated and imperceptibly working its effect, through an incalculable series of generations' *(Natural, Theology,* p. 156). He satirized the very possibility of evolution in a passage whose mocking tone now rebounds on itself:

> A piece of animated matter, for example, that was endued with a propensity to *fly,* though ever so shapeless, though no other we will suppose than a round ball, to begin with, would, in a course of ages, if not in a million of years, perhaps in a hundred millions of years (for our theorists, having eternity to dispose of, are never sparing in time), acquire *wings.* The same tendency to loco-motion in an aquatic animal, or rather in an animated lump which might happen to be surrounded by water, would end in the production of *fins:* in a living substance, confined to the solid earth, would put out *legs* and *feet;* or if it took a different turn, would break the body into ringlets, and conclude by *crawling* upon the ground. *(Natural Theology,* pp. 463–4)[57]

There was in the evolutionary discourse of the time, therefore, a specific challenge to Darwin to demonstrate the place of congenital deformities in nature. The view that providence was always aiming at some pre-ordained idea of perfection, or even the idea that nature permitted variations only within certain limits, invalidated Natural Selection because it would never allow for major changes. An ideal of physical perfection for a species must imply that species in general were originally designed and never changed. Darwin answered this challenge by showing that what were termed 'monstrosities'

---

[57] George Canning's riposte to Erasmus Darwin's *The Loves of the Plants* – the 1798 *Loves of the Triangles* – presented a satirical view of the same process, where vegetables 'by degrees detached themselves from the surface of the earth, and supplied themselves with wings or feet ... [Others] by a stronger effort of *volition,* would become men. These, in time, would restrict themselves to the use of their *hind* feet; their *tails* would gradually rub off, by sitting in their caves or huts,' (Quoted by Desmond King-Hele, *Erasmus Darwin and the Romantic Poets* (Basingstoke: Macmillan, 1986) p. 140).

('some considerable deviation of structure in one part, either injurious or not useful to the species') were in some cases no different from 'variations', ('what are called monstrosities ... graduate into varieties'), and could in fact be inherited:

> Some authors use the term 'variation' in a technical sense, as implying a modification directly due to the physical conditions of life; and 'variations' in this sense are supposed not to be inherited: but who can say that the dwarfed condition of shells in the brackish waters of the Atlantic, or dwarfed plants on Alpine summits, or the thicker fur of an animal from far northwards, would not in some cases be inherited for at least some few generations? *(The Origin,* p. 101)

Darwin showed that such 'peculiarities' (or, as he also named them, 'deviations', 'individual differences', 'aberrant species' and 'divergences') were in fact the raw material for evolution, and that these apparent anomalies could be the first steps towards new species which were better adapted to their environment than were their predecessors. Whereas in Natural Theology variations on 'perfection' were aberrations which needed to be explained away, in Darwin's theory they were crucially necessary to the whole system. The Darwinian thesis thus foregrounded and valued physical diversity, individuality, difference – even deviance – as a means of change and development, in a way which recalls arguments deployed in Romantic contexts and particularly by Byron in both political and personal arenas.

Byron's response to his own physical deformity (congenital club foot arises from a dominant gene with variable expressivity, caused by a mutation) sometimes took the form of a defiance which claimed that there is a special value in human bodies that are *not* made in the perfect image of God.[58] Arnold in *The Deformed* states what, from the evidence of Byron's letters, was also a personal creed for Byron:

> Deformity is daring.
> It is its essence to o'ertake mankind
> By heart and soul, and make itself the equal –
> Aye, the superior of the rest. There is
> A spur in its halt movements, to become
> All that the others cannot, in such things
> As still are free to both, to compensate
> For stepdame Nature's avarice at first.
> They woo with fearless deeds the smiles of fortune,
> And oft, like Timour the lame Tartar, win them.
> *(Deformed Transformed,* I. i. 313–22)[59]

---

[58] See Theodosius Dobzhansky, *Mankind Evolving: The Evolution of the Human Species* (New Haven: Yale University Press, 1962) p. 108. Byron was acutely anxious lest his daughter Ada should inherit his deformity: 'On the second day after the birth of the child Lord B finding me only in the room with the Infant asked particularly about its feet, and upon my offer of showing them to him said thank you, thank you'. Mary Anne Clermont, quoted by Doris Langley Moore in *Ada, Countess of Lovelace* (London: John Murray, 1977) p. 18.

[59] Byron echoes and reclaims Sir Francis Bacon's argument, in 'On Deformity': see *The Essayes or Counsels, Civill and Morall,* ed by Michael Kiernan, (Cambridge, Mass.: Harvard

Both Alain-René Lesage's satiric and Mephistophelian devil Asmodeé from whom Byron chose his own sobriquet *Le Diable Boiteux*, and the Stranger/Caesar in *The Deformed Transformed*, deliberately select deformed bodies in which to operate, rejecting the options of the perfect and beautiful forms available to them.[60] Byron's marginal note to the unfinished *Deformed Transformed* indicates that he planned to develop the plot by having Arnold become jealous of Caesar because his wife Olimpia seems to prefer the hunchback ('owing to the power of intellect &c. &c. &c.') to Arnold in his beautiful (and supernaturally-sustained) Achillean shape. In the Pope/Bowles controversy, Byron sprang to the defence of the hunchbacked Alexander Pope for his supposed treatment of Lady Mary Wortley Montagu, and protested that the poet's deformity need not have made him unattractive to men or women, producing several anecdotes and pieces of evidence about the passionate nature and sexual attractiveness of hunchbacks and other deformed people *(BCPW*, pp. 169–70).[61] His reaction to attacks on himself in the Tory press in 1814 which called him 'a sort of R[ichar]d 3d. — deformed in mind & *body*' (written, Byron believed, by 'Rosa Matilda': Charlotte Dacre, wife or mistress of the editor of the *Morning Post)* was to proclaim in Greek in his letters (though not in print) 'a lame man fucks best' *(BLJ*, IV. 49–51).[62]

Sexual selection is, of course, a vital feature of the natural selection which leads to evolution: 'This depends, not on a struggle for existence, but on a

---

University Press, 1985) p. 133: 'Whosoever hath any Thing fixed in his Person, that doth enduce Contempt, hath also a perpetuall Spurre in himselfe, to rescue and deliver himselfe from Scorne: Therefore all *Deformed Persons* are extreme Bold. First, as in their own Defence, as being exposed to Scorn; But in Processe of Time, by a Generall Habit.' He also follows Pliny, at the opening of Book VII of the *Natural History,* or Montaigne, in the 'Apology for Raymond Sebond', in characterizing nature as a stepmother, rather than a mother, to humankind.

[60] For Byron's references to himself as *Le Diable Boiteux* see *BLJ*, IV. 50 and 51 and *BLJ*, X. 136. Lesage describes Asmodée as 'leaning on two crutches ... he limped on the legs of he-goat, with a long face, a pointed chin, a black and yellow complexion and a very squashed nose'. Asmodée claims: 'I could have shown myself to your eyes clothed in a beautiful fantastic body. Nevertheless, although I am so crippled, I can get about very well; you will see how agile I am.' *Le Diable Boiteux*, pp. 88–90 (my translations).

[61] Byron follows Pope himself in this, who wrote to Lady Mary in 1716, as she was about to depart for the East: 'I am capable myself of following one I lov'd, not only to Constantinople, but to those parts of India, where they tell us the Women like best the Ugliest fellows, as the most admirable productions of nature, and look upon Deformities as the Signatures of divine Favour': *The Correspondence of Alexander Pope,* ed by George Sherburn, 5 vols (Oxford: Clarendon Press, 1956) I. 364.

[62] Byron's choice of Cain as a subject reflects the links between this figure, deformity and animals which have been part of the characterization of Cain since the earliest times. Ruth Mellinkoff points out in *The Mark of Cain* (Berkeley: University of California Press, 1981) p. 59, how 'the idea of deformity or disfigurement [in 'the mark of Cain'] was exaggerated to accentuate the beastlike nature of Cain. Cain was born entirely human. Not a daemon, demihuman, or monster. But his rapid descent into evil, his lack of natural feeling so grotesquely displayed in the brother-murder, provoked interpretations that emphasised the bestial quality of his character. Motifs or themes blurring the edges between Cain seen as a beastlike man, and Cain seen as a manlike beast, appeared among the many legends enlarging the biblical story. A partial metamorphosis of Cain, that is, from depraved man into a half human, almost half animal creature is hinted at [in the idea that] Cain was marked with a horn.'

struggle between the males for possession of the females: the result is not death to the unsuccessful competitor, but few or no offspring' *(The Origin,* p. 136). Direct physical combat is not, however, the only means through which males struggle to be selected by mates: amongst birds, for instance, 'the severest rivalry between the males of many species [is] to attract by singing the females' *(The Origin,* p. 137). Darwin also specifies 'peculiarities' of appearance amongst animals and birds which seem to attract females, providing the males with 'charms' which give them some advantage in the same way that specific means of defence or weapons do, including 'the tuft of hair on the breast of the turkey-cock' which 'can hardly be either useful or ornamental to this bird' *(The Origin,* p. 138). Paley also has recourse to anatomical oddities for his argument in the opposite direction: instancing the air bladder of fish, the fang of the viper, the bag of the opossum, the camel's stomach and the woodpecker's tongue as examples of strange features which ensure survival in hostile surroundings: they could, he says, not possibly have arisen from evolution and could only be the product of design *(Natural Theology,* pp. 264–71).

Similarly, the Stranger's sarcastic description of Arnold's deformities in Part I of *The Deformed Transformed* draws attention to their animal-like features in a way which makes use of the Paleyian argument, alluding to the notion of the potential 'usefulness' of apparent deformities, and at the same time de-centres man from the world in favour of animals:

ARNOLD   Do you – dare *you*
         To taunt me with my born deformity?

STRANGER   Were I to taunt a buffalo with this
         Cloven foot of thine, or the swift dromedary
         With thy sublime of humps, the animals
         Would revel in the compliment. And yet
         Both beings are more swift, more strong, more mighty
         In action and endurance than thyself,
         And all the fierce and the fair of the same kind
         With thee. Thy form is natural: 'twas only
         Nature's mistaken largess to bestow
         The gifts which are of others upon man.

ARNOLD Give me the strength then of the buffalo's foot,
         When he spurns high the dust, beholding his
         Near enemy; or let me have the long
         And patient suffering of the desart-ship,
         The helm-less dromedary; – and I'll bear
         Thy fiendish sarcasm with a saintly patience.

STRANGER   I will.
         (Deformed Transformed, I. i. 101–19)[63]

---

63   Byron also seems to echo Pope's *Essay on Man,* I. 175–88:
        Now looking downwards, just as grieved appears,
        To want the strength of bulls, the fur of bears.
        Made for his use all creatures if he call,
        Say, what their use, had he the power of all?

The return of the dead

This chapter has aimed to add to the existing body of work on the currency of evolutionary ideas in imaginative work in the half-century before *The Origin of Species* by demonstrating how the scientific and theological evolutionary debate of the early nineteenth century was quarried as a source of material by imaginative writers in this area. Byron, Keats and Tennyson use and foreshadow concepts that were at the core of Darwin's work, so that Darwin seems at times to demonstrate a benign version of Harold Bloom's *Apophrades,* or 'The Return of the Dead', whereby strong writers 'achieve a style that captures and oddly retains priority over their precursors' making one 'believe, for startled moments, that they are being *imitated by their ancestors'* (Bloom's italics).[64] Turning to Bloom's more orthodox version of the theory of influence, I shall consider briefly to what extent in his scientific writing Darwin may have been influenced (and been made anxious in a Bloomian way) by his youthful reading of Romantic poetry.

'Up to the age of thirty, or beyond it,' Darwin claimed in his *Autobiography* (p. 158), 'poetry of many kinds, such as the works of Milton, Gray, Byron, Wordsworth, Coleridge and Shelley, gave me great pleasure'.[65] Deep and detailed readings and misreadings of *Paradise Lost* are certainly common to Darwin, Keats, Byron and Tennyson, in the sense that Milton's version of the myth of the Creation and Fall is the Bloomian precursor with whose influence all four grapple, in the works I have discussed. If Darwin was influenced by his reading of Coleridge it seems also to have been mainly in this, 'reactive', sense, since Coleridge's scientific theories (which drew heavily on Friedrich Schelling, Lorenz Oken and *Naturphilosophie)* pointed in an opposite direction to Darwinism. A reading of Wordsworth's verse may have helped to direct Darwin towards the minute study of natural objects, but the main statement of intent in the Preface to *The Excursion* (commented on already in Chapter Five) seems to run directly counter to Darwin's purpose in *The Origin:*

> my voice proclaims
> How exquisitely the individual Mind
> (And the progressive powers perhaps no less
> Of the whole species) to the external World
> Is fitted – how exquisitely too –

> Nature to these, without profusion, kind,
> The proper organs, proper powers assign'd;
> Each seeming what compensated of course,
> Here with degrees of swiftness, there of force ...
> Each beast, each insect, happy in its own:
> Is Heaven unkind to man, and man alone?
> Shall he alone, whom rational we call,
> Be pleased with nothing, if not blest with all?

[64] Harold Bloom, *The Anxiety of Influence: A Theory of Poetry,* 2nd edn (London: Oxford University Press, 1997) p. 141.

[65] Darwin's notebooks also allude to 'Byron': but in fact the index to Darwin's collected *Works* mistakenly allocates these references to the poet when they are actually to the writings of two of his seafaring relatives: his grandfather, Admiral John Byron, and his cousin, later the seventh baron, Captain George Anson Byron.

Theme this but little heard of among men –
The external world is fitted to the Mind;
And the creation (by no lower name
Can it be called) which they with blended might
Accomplish.
(Excursion, Preface, 62–70)

Darwin did not endorse Cuverian Catastrophism – the evolutionary theory which had most appeal for Byron's imagination. Darwinism required only a very long time-scale for the steady development of separate species which it postulated, rather than any sudden destructions of the environment, combined with mutations or 'leaps' in species change: 'Natura non facit saltem' was one of the few principles on which Darwin and Paley agreed. The prevailing geological thesis on which *The Origin* was based was Charles Lyell's version of Uniformitarianism, and in fact scientists now take more seriously than Darwin did the kind of Catastrophe (and humankind's response to it) which Byron envisaged in conversation with Medwin in 1823:

> 'Who knows whether, when a comet shall approach this globe to destroy it, as it often has been and will be destroyed, men will not tear rocks from their foundations by means of steam, and hurl mountains, as the giants are said to have done, against the flaming mass? – and then we shall have the traditions of the Titans again, and wars with Heaven.'
> (Medwin's *Conversations,* p. 188)[66]

Byron's resistance to any certainty about the superiority of humanity to other species, and his preoccupation with the apparent unjust anomaly of a cruel and vindictive God in a human-centred world, challenge religious and scientific orthodoxies in areas which would have been of evident interest to the young Darwin. If, as I have suggested, Byron and Darwin both took the answers of Natural Theology and turned them into questions, then Byron's imaginative envisioning of the questions was clearly available to the young Darwin at the time he began to define the topics he wanted to tackle.

Byron's imaginative response left the questions *as* questions: he drew attention to anomalies, inconsistencies, and failures of logic in Natural Theology's conception of the world, but he formulated no reply in philosophical or other terms. Darwin posed the same questions, but he also created alternative answers, with far-reaching effects. These answers do not propose a more consoling world for humankind than Byron's questions do: they endorse the minimization of man in the universe and, although they obviate the need for Byron's cruel and unjust Jehovah, they propose instead a blind chance which is indifferent to the development or the fate of humankind. Byron also questioned Natural Theology through the ironic fictions he span around the theories of the orthodox Churchmen who used Cuvier's work to find an apparently perfect fit between the geological and biblical messages. Despite the conservatism of the evolutionary theories he was taught during his medical training, Keats also evidently perceived the way in which evolution

---

[66] The Spaceguard Foundation, established in 1996, involves scientists from around the world in an asteroid surveillance programme and two NASA working groups have been established to consider the problems posed by 'near earth objects'.

could work against the status quo, and was aware of the Church's attempt in
the first decades of the century to capture geological inquiry for its own ends.
'Geology, although ... bearing on the moral and religious belief of mankind,
will certainly never prosper,' commented a radical journalist twenty years later,

> as long as, first the science itself, and even scientific specimens, will be
> a matter of traffic and trade amongst the savants, and the higher classes
> in general – and, second, as long as geologists will so much depend on
> right reverends, right honourables, &c., in fine, on those who are
> interested in keeping up the usual common-place *go* in society [i.e. the
> status quo].[67]

The irony of Byron's fictions was misunderstood by critics such as Francis
Jeffrey and by friends including Thomas Moore – partly because Byron had
been teasingly obfuscatory about his polemical and satirical intentions.[68]
Darwin by contrast made every effort to be conciliatory, clear and methodical
in putting forward his arguments and proofs, but he was nevertheless criticized
by scientists such as Alfred Wallace (his fellow presenter on evolution to the
Linnaean Society) for endowing natural forces with human attributes. Where
Byron using fiction was mistaken to be literal, Darwin in trying to be as
precisely literal as possible was wrong-footed for using fiction and metaphor in
a way which risked undermining the message about the impersonal and non-
human-centred nature of the processes he was attempting to define. Darwin
tried to distance himself from the literary and theological language and
references which had accrued around the scientific study of evolution by
deploying similes drawn from other areas, including a famous one which
shows how different species compete for the earth's resources: 'The face of
Nature may be compared to a yielding surface, with ten thousand sharp wedges
packed close together and driven inwards by incessant blows, sometimes one
wedge being struck and then another with greater force'*(The Origin,* p. 119).
This is still an image of human hands at work, however. Darwin was being
creatively 'figurative' in a way he may have learnt from the imaginative work
he read in his youth, and he was obliged to defend himself against the charge
of having personified 'natural selection':

> For brevity [sic] sake I sometimes speak of Natural Selection as an
> intelligent power; in the same way as astronomers speak of the
> attraction of gravity as ruling the movements of the planets. ... I have,
> also, often personified the word nature; for I have found it difficult to

---

[67] William Chilton, 'Geological Revelations', in *Oracle,* 29 August 1843, quoted in
Desmond, 'Artisan Resistance', pp. 89–90.

[68] Francis Jeffrey complained in his review of *Cain* in the *Edinburgh Review* in 1822: 'It is
but a poor and pedantic sort of poetry that seeks to embody nothing but metaphysical subtleties
and abstract deductions of reason. ... We therefore think that poets ought fairly to be confined
to the established creed and morality of their country, or to the *actual* passions and sentiments
of mankind ... There is no poetical road to metaphysics' (Rutherford, *Critical Heritage,*
pp. 234–5).

avoid this ambiguity; but I mean by nature only the aggregate notion and product of many natural laws – and by laws only the ascertained sequence of events.[69]

Darwin's work amply fulfilled Wordsworth's prediction that 'the labours of men of Science' would one day create a 'material revolution, direct or indirect, in our condition, and in the impressions which we habitually receive'. Wordsworth's hope, however, that poets would be available to 'carry sensation into the midst of the objects of the Science itself' and his wish to 'familiarize' science to men, making it 'ready to put on ... a form of flesh and blood', characterize a way of approaching science which was beginning to be judged inappropriate by the middle of the nineteenth century. The effect of Darwin's message about the universe was to undermine the designed-for-man arguments of Natural Theology, and to make metaphors which figured the universe in human terms seem unwarranted. Darwin's struggle to de-centre man from the universe included a struggle with human-centred language, and involved the repudiation of Wordsworth's hope that a 'transfigured' Science would in any way wish to make itself welcome 'as a dear and genuine inmate of the household of man'.

---

[69] Charles Darwin, *Variation of Animals and Plants under Domestication,* ed by Paul H. Barrett and R. B. Freeman (London: Pickering, 1988) I. 5–6.

In Conclusion

Animals Then and Now

Wordsworth's comparison of the poet and the 'man of science', discussed in the previous chapter, was part of a larger project to bring the poet into relationship with a group of highly-educated professionals – what Coleridge later called 'the Clerisy' – which, the two poets hoped, would form the new heart of the country and the leaders of the state.[1] So although Wordsworth evokes poetry in terms of the sublime as 'the breath and finer spirit of all knowledge,' and the poet as 'the rock of defence of human nature', in more everyday language he also compares the poet's authority and social standing with that of historians, lawyers, physicians, astronomers and 'natural philosophers'; and this argument enables him both to resituate the poet in a respected place within society, and also to demonstrate the particular, unique, offering which he believed the poet could make to that society.

The broadly historicist method of this study has followed a similar path. By considering professional or cultural discourses which include approaches to animals alongside selfconsciously-literary or poetic writing of the Romantic period, it has aimed both to define what were the prevailing views about this topic in this place and time, and also – without necessarily endorsing Wordsworth's exalted definition of the poet – to consider work of the period which makes deliberate use of imaginative features and linguistic art forms (particularly verse) in its approaches to animals.[2] These concluding remarks seek to summarize the outcomes of this process.

'Prevailing' is of course a highly-charged word in cultural terms. Raymond Williams commented that 'the only sure fact about the organic society is that it has always gone'; and the identification of a specific 'Romantic Ideology' by Jerome McGann and others is only one of many critical developments of the last quarter-century that have enabled us to see that Romanticism, in the sense of a prevailing culture which organised society, is not only 'gone' but in that sense never really existed.[3] Williams's distinction between the 'residual', 'dominant' and 'emergent' aspects of Renaissance culture has helped to undermine the concept of a single 'spirit of the age' (with which, in its invocation by Virginia Woolf, I opened this study) and to replace it with a more dynamic model which has freed a space for the presentation of a more

---

[1] Coleridge, *Church and State*, p. 46: 'The CLERISY of the nation, or national church, in its primary acceptation and original intention comprehended the learned of all denominations; – the sages and professors of the law and jurisprudence; of medicine and physiology; of music; of military and civil architecture; of the physical sciences; ... in short, all the so called liberal arts and sciences, the possession and application of which constitute the civilization of a country, as well as the Theological.'

[2] A high level of linguistic and rhetorical skill is, of course, often evident in the writing of the period which does not view itself as specifically 'literary' (children's fiction, parliamentary debates and Darwin's *Origin of Species* are all cases in point).

[3] Raymond Williams, *The Country and the City* (London: Hogarth, 1973) pp. 9–12.

complex picture of a culture, including its subversive and marginalized elements.[4]

Williams's threefold characterisation of culture can be usefully applied to concepts about animals in the Romantic period, and such an approach also helps to counter the habit of cherry-picking only the ideas from that era which have gone on to become important in our own, and thus ignoring the full picture. Some of the most interesting ideas about animals current in the Romantic period were the 'residual' ones: those which survived from earlier times, but were already subject to challenge or being forgotten then, and are no longer recognized now. These include the doctrine of signatures, which caused many at the end of the eighteenth century still to believe that eating bull's flesh (or even participating in bull-baiting or bull-fighting) would give people bull-like characteristics, and was therefore especially valuable during war time. The obverse of this belief was the Shelleyan expectation that renouncing meat-eating would by physiological means alone render people less bellicose, calmer and more humane: an idea which perhaps still has some currency today, although Adolf Hitler's vegetarianism strains the theory somewhat.

The eighteenth-century idea that showing humanity to animals is merely 'practice' – for children especially – for the real business of being kind to one's fellow human beings is, as I have shown, traceable back to Pythagoras via Locke, Aquinas, and others. It was a way of thought that was being replaced in the Romantic period by the concept that being kind to animals was worthwhile in its own right, and it is no longer one of the reasons we articulate when we teach children to look after their pets (although it may be residually present, all the same). Nor do we now tell each other that being kind to animals will have the specific advantage of making us less like animals, or ban butchers from juries on the grounds that killing animals will correlate with inhumanity towards human beings.

I have indicated how Byron's theriophily was drawn from a classical and Renaissance tradition of scepticism which was considered unacceptably out-of-date even in his own time. It was also suspect in orthodox circles because it related to French philosophical scepticism, and thus was associated with Jacobinism, and, as a result of this, Byron's work written in this spirit was widely misread as simply misanthropic. Another 'old-fashioned' or residual idea about animals which still had unexpected force in the Romantic period was the preoccupation with whether or not beasts had souls. As I have shown, this classically-founded question had reached the moment of its most intense debate in seventeenth-century Cartesianism, but it was still of considerable importance to much later writers including Cowper, Trimmer, Wordsworth, Coleridge, Byron and Shelley.

My study has aimed to show that animal-related topics had a new or emergent (as well as an old) importance in the Romantic period. Questions about liberty, equality, fraternity, political franchise and the pursuit of happiness, raised in the context of the American and French Revolutions,

---

[4] Raymond Williams, *Marxism and Literature* (Oxford: Oxford University Press, 1977) p. 91. E. M. W. Tillyard's *The Elizabethan World Picture* (1943) was the particular target of Williams's critique, but the concept of 'the spirit of the age' dates back at least to Hazlitt (1825).

were brought to bear upon not only the relationships between different human groups, but also those between human beings and other animals. After its first use in the late eighteenth century in association with liberty and rights for slaves, women and other oppressed groups, the link between animals and the 'lower' ranks of humankind was deployed by conservative as well as politically-radical writers, so that one sees Trimmer using animals as a means of carefully teaching children the distinctions of social class, just as confidently as Bentham was making them illustrate the need to disregard physical features such as skin colour in the allocation of rights and liberty.

A major area of animal-related concern that was dominant in the Romantic period was the question of the existence of a beneficent Providence, to which were tied questions about the special creation of Man, humankind's centrality to the universe, the biological relationship between humans and animals, and their possible descent from a common ancestor. Sigmund Freud (in 1917) defined 'the researches of Charles Darwin and his collaborators and forerunners' as one of the three severe 'blows' or 'wounds' that 'the universal narcissism of men, their self-love', had suffered from science, summarizing the process as one in which the Darwinite researches had 'put an end to this presumption on the part of man'.[5] One of the major manifestations of this 'presumption' in the period and country of Darwin's 'forerunners' was in an anxiously reactive form, as an over-emphatic insistence on the special, divinely-appointed, place of Man in the world. God's benevolence to humankind in this respect was emphasised in the Natural Theology of writers such as William Paley and in the Bridgewater Treatises which were intended, by the eponymous Earl who left the money in his will to establish them, to demonstrate 'the Power, Wisdom, and Goodness of God, as manifested in the Creation'. This project included a philosophical and theological struggle with the difficult issues of the existence of pain and disease and of animals' apparent cruelty towards each other, which were also of particular concern to those writing for children. Wordsworth's perception of the need for the poet to be a 'rock of defence of human nature' and his reliance on a Nature which 'never did betray / the heart that loved her' can be read as a deistic or secular version of the same conviction that the natural world, including animals, was beneficently designed for human beings to live in.

The idea that being kind to animals could make humankind more human(e) and less 'brutal' was, as discussed, an ancient idea under modification in the Romantic period by the perception that kindness could be an end in itself. At the beginning of the period, the simple exercise of kindness was capable for a while of disguising the difference in stance between those who wanted to be kind to animals in order to emphasise their own difference from the beasts, and those who wanted to treat beasts well because of animals' similarity to humankind. Such important distinctions between liking and likeness, kindness and kinship could not remain submerged for long, however, although the old

5   From 'A Difficulty in the Path of Psycho-Analysis', in *The Standard Edition of the Complete Psychological Works of Sigmund Freud*, ed by James Strachey, 24 vols (London: Hogarth Press and Institute of Psycho-Analysis, 1953–74) XVII. 140–1 (see Beer, *Darwin's Plots*, p. 12). The other two blows to human self-esteem identified by Freud were the 'cosmological', associated with Copernican theory, and the 'psychological', associated with psychoanalysis.

notion was given an unexpected new lease of life, early in the nineteenth
century, by the supposedly scientific 'discovery' that human embryos passed
through stages corresponding to those of the 'lower' animals before reaching
their final shape in humankind, and thus that becoming a fully developed
human meant repudiating the animal in oneself. This belief may have grown
up in opposition to Rousseau's novel expression, in the early part of the period,
of ideas about the general superiority of 'natural' states and behaviour over
those that were cultivated or civilized. The Rousseauian plan for the
upbringing of children was often encapsulated in animal metaphors which
related it to the notion (prominent in Buffon) that animal species 'degenerated'
under domestication, and the 'Pythagorean' dietetic regime practised by
Shelley and others at the turn of the eighteenth and nineteenth centuries drew
heavily on the idea that humankind was not 'naturally' formed to eat meat and
that doing so led to physical and moral 'degeneracy'.

   This form of vegetarianism *avant la lettre* leads on to the consideration of
areas of thought and behaviour concerning animals which were marginal or
emergent in the Romantic period but have become powerful, and in some cases
central, in Western culture since then. Modern vegetarianism in the West is
often motivated by simple abhorrence of rearing and killing animals to eat, but
a fear of the adverse physiological and psychological effects of meat-eating,
similar to that current in the Romantic period, has in our own time gained force
from the BSE and other crises in meat-production. Concepts about the
economy of nature were popularized by Linnaeus and others in the mid-
eighteenth century, but the idea that human activities could harm this
economy; that the landscape and 'the environment' were at risk from the
effects of civilization rather than benefiting from them, and therefore that
humankind might have a crucial role to play in preserving the natural world,
first emerged to consciousness in the Romantic period. In this form it is a
precursor of powerful modern ecological movements, and Wordsworth in
particular has been associated with its first stirrings.

   Finally, the idea of a consubstantiality or confraternity between human
beings and animals, which underpins much current thinking about human
behaviour and adds greatly to the value placed on all animals, especially wild
ones, is one which had its beginning in the late-eighteenth and early-nineteenth
centuries. In a metaphorical, spiritual or feeling-based form, allied to many
different currents of contemporary thought, including eighteenth-century
sympathetic sensibility and revolutionary religious scepticism, this perception
of human/animal kinship in the Romantic period foreshadows what Darwinism
demonstrated half a century later through biological means. In my view, this is
the most important of the animal-related concepts for which the Romantic
period provides a precursor for the opinions of the Western world at the start of
the twenty-first century.

Bibliography

Abbreviated titles

BLJ   Byron, George Gordon Noel, Lord, *Byron's Letters and Journals,* ed by Leslie A. Marchand, 13 vols (London: John Murray, 1973–94)

LBCMP   Byron, George Gordon Noel, Lord, *Lord Byron: The Complete Miscellaneous Prose* ed by Andrew Nicholson (Oxford: Clarendon Press, 1991)

LBCPW   Byron, George Gordon Noel, Lord, *Lord Byron: The Complete Poetical Works,* ed by Jerome J. McGann, 7 vols (Oxford: Clarendon Press, 1980–93)

WLBLJ   Byron, George Gordon Noel, *Letters and Journals,* ed by Rowland E. Prothero, 6 vols (London: John Murray, 1902–04)

WLBP   Byron, George Gordon Noel, Lord, *The Works of Lord Byron: Poetry,* ed by Ernest Hartley Coleridge, 7 vols (London: John Murray, 1898–1904)

Primary sources cited

Aikin, John, and Anna Laetitia Barbauld, *Evenings at Home; or, The Juvenile Budget Opened,* 3 vols (London: J. Johnson, 1792–96)

Amyot, Thomas, *Speeches in Parliament of the Right Honourable William Windham; To which is Prefixed some Account of his Life* (London: Longman etc., 1812)

Anon., *Goody Two-Shoes: A Facsimile Reproduction of the Edition of 1766,* ed by Charles Welsh (London: Griffith & Farran, 1881)

Anti-Jacobin; or, Weekly Examiner, January 1798 (London: J. Wright, 1798)

Aquinas, Thomas, *Summa Theologiae: A Concise Translation* ed by Timothy McDermott (London: Eyre & Spottiswoode, 1989)

Austen, Jane, *Northanger Abbey* ed by Anne Henry Ehrenpreis (Harmondsworth: Penguin, 1972)

Bacon, Sir Francis, *The Essayes or Counsels, Civill and Morall,* ed by Michael Kiernan (Cambridge, Mass.: Harvard University Press, 1985)

Barbauld, Anna Laetitia, *The Works of Anna Laetitia Barbauld,* ed by Lucy Aikin, 2 vols (London: Longman, 1825)

Bayle, Pierre, *Mr Bayle's Historical and Critical Dictionary: The Second Edition; to which is Prefixed the Life of the Author, Revised, Corrected and Enlarged by Mr Des Maizeaux,* 5 vols (London: Knapton., 1734)

Beckford, William, *Vathek,* ed by Roger Lonsdale (Oxford: Oxford University Press, 1983)

Bentham, Jeremy, *The Collected Works of Jeremy Bentham: An Introduction to the Principles of Morals and Legislation* ed by J. H. Burns and H. L. A. Hart (London: University of London and Athlone Press, 1970)

Bewick, Thomas, [and Ralph Beilby], *A General History of Quadrupeds; The Figures Engraved on Wood by T. Bewick* (Newcastle-upon-Tyne: Beilby and Bewick, 1790)

Blessington, Marguerite, Countess of, *Lady Blessington's Conversations of Lord Byron,* ed by Ernest J. Lovell Jr (Princeton: Princeton University Press, 1969)

Blake, William, *The Complete Graphic Works of William Blake,* compiled by David Bindman and Deirdre Toomey (London: Thames and Hudson, 1978)

Blake, William, *The Complete Poems of William Blake,* ed by Alicia Ostriker (Harmondsworth: Penguin, 1977)

Blake, William, *Selected Poems of William Blake,* ed by F. W. Bateson (London: Heineman, 1957, repr. 1981)

Boswell, James, *Boswell's Life of Johnson* ed by R. W. Chapman (Oxford: Oxford University Press, 1985)

Browning, Robert, *The Complete Works of Robert Browning* ed by John C. Berkey, Allan C. Dooley and Susan C. Dooley, 9 vols (Athens, Ohio: Ohio University Press, 1996)

Browning, Robert, *The Poems of Browning,* ed by John Woolford and Danny Karlin (Harlow: Longman, 1991)

Buckland, William, *Vindiciae Geoligicae; or, The Connexion of Geology with Religion Explained* (Oxford: Oxford University Press, 1820)

Buffon, Georges-Louis Leclerc, comte de, *Histoire naturelle, générale et particulière,* 44 vols (Paris: Imprimerie Royale, puis Plassan, 1749–1804)

Burke, Edmund, *A Philosophical Enquiry into the Origin of our Ideas of the Sublime and Beautiful,* ed by J. T. Boulton (London: Routledge & Kegan Paul, 1958)

Burke, Edmund, *Reflections on the Revolution in France,* ed by A. J. Grieve (London: Everyman, Dent, 1967)

Burns, Robert, *The Poetical Works of Burns,* ed by Raymond Bentman (Boston: Houghton Mifflin, 1974)

Cambridge Union Society, The, *Laws and Transactions of the Union Society* (Cambridge, 1830)

Carr, Sir John, *Descriptive Travels in the Southern and Eastern Parts of Spain and the Balearic Isles, in the Year 1809* (London: Sherwood, Neely and Jones, 1811)

Chalmers, Thomas, *The Evidence and Authority of the Christian Revelation* (Edinburgh: Blackwood, 1814)

Chalmers, Thomas, 'On Natural Theology,' vols I and II of *The Works of Thomas Chalmers,* 25 vols (Glasgow: Collins, 1836–42)

Chambers, Robert, *Vestiges of the Natural History of Creation,* ed by Gavin de Beer (Leicester: Leicester University Press, 1969)

Clare, John, *John Clare,* ed by Eric Robinson and David Powell (Oxford: Oxford University Press, 1984)

Cobbett, William, *Cottage Economy* (London: C. Clement, 1822)

Coleridge, Samuel Taylor, *Biographia Literaria; or, Biographical Sketches of my Literary Life and Opinions,* ed by James Engell and W. Jackson Bate, 2 vols, *The Collected Works of Samuel Taylor Coleridge,* Bollingen Series (London: Routledge & Kegan Paul, 1983)

Coleridge, Samuel Taylor, *Coleridge's Shakespearean Criticism,* ed by Thomas Middleton Raysor, 2 vols (London: Constable, 1930)

Coleridge, Samuel Taylor, *Collected Letters of Samuel Taylor Coleridge* ed by Earl Leslie Griggs, 6 vols (Oxford: Oxford University Press, 1956–57)

Coleridge, Samuel Taylor, *Lay Sermons* ed by R. J. White, *The Collected Works of Samuel Taylor Coleridge,* Bollingen Series (London: Routledge & Kegan Paul, 1972)

Coleridge, Samuel Taylor, *Lectures 1795 on Politics and Religion,* ed by Lewis Patton and Peter Mann, *The Collected Works of Samuel Taylor Coleridge,* Bollingen Series (London: Routledge & Kegan Paul, 1971)

Coleridge, Samuel Taylor, *On the Constitution of the Church and State,* ed by John Colmer, *The Collected Works of Samuel Taylor Coleridge,* Bollingen Series (London: Routledge & Kegan Paul, 1976)

Coleridge, Samuel Taylor, *The Notebooks of Samuel Taylor Coleridge,* ed by Kathleen Coburn, 4 vols (London: Routledge & Kegan Paul, 1957–90)

Coleridge, Samuel Taylor, *A Critical Edition of the Major Works,* ed by H. J. Jackson (Oxford: Oxford University Press, 1985)

Coleridge, Samuel Taylor, *Poetical Works*, ed by E. H. Coleridge, 2 vols (Oxford: Clarendon Press, 1912)

Coleridge, Samuel Taylor, *Table Talk: Recorded by Henry Nelson Coleridge (and John Taylor Coleridge)* vol. I, ed by Carl Woodring, *The Collected Works of Samuel Taylor Coleridge*, Bollingen Series (London: Routledge, 1990)

Coleridge, Samuel Taylor, *The Watchman*, ed by Lewis Patton, *The Collected Works of Samuel Taylor Coleridge*, Bollingen Series (London: Routledge & Kegan Paul, 1970)

[Coventry, Frances], *History of Pompey the Little; or, The Life and Adventures of a Lap-dog* (London, 1751)

Cowper, William, *The Poems of William Cowper*, ed by John H. Baird and Charles Ryskamp, 3 vols (Oxford: Clarendon Press, 1980–95)

Cuvier, Georges, baron, *Essay of the Theory of the Earth Translated from the French of M. Cuvier ... by Robert Kerr ... with Mineralogical Notes and an Account of Cuvier's Geological Discoveries by Professor Jameson* (Edinburgh: William Blackwood and John Murray, 1813)

Cuvier, Georges, baron, *Recherches sur les ossemens fossiles de quadrupèdes*, 4 vols (Paris, 1812)

Darwin, Charles, *Darwin's Early and Unpublished Notebooks*, ed by Paul H. Barrett (London: Wildwood House, 1974)

Darwin, Charles, *Foundations of the Origin of Species*, ed by Francis Darwin (1909), repr. Pickering Masters: *The Works of Charles Darwin*, vol. 10 (London: Pickering, 1986)

Darwin, Charles, *The Autobiography of Charles Darwin* ed by Nora Barlow, Pickering Masters: *The Works of Charles Darwin*, vol. 29 (London: Pickering, 1989)

Darwin, Charles, *The Correspondence of Charles Darwin*, ed by Frederick Burkhardt and others, 8 vols (Cambridge: Cambridge University Press, 1985–93)

Darwin, Charles, *The Origin of Species by Means of Natural Selection; or, The Preservation of Favoured Races in the Struggle for Life, 1st edn, 1859*, ed by J. W. Burrow (Harmondsworth: Penguin, 1968)

Darwin, Charles, *The Life and Letters of Charles Darwin; Including an Autobiographical Chapter*, ed by Francis Darwin, 3 vols (London: John Murray, 1888)

Darwin, Charles, *Variation of Animals and Plants under Domestication*, ed by Paul H. Barrett and R. B. Freeman, Pickering Masters: *The Works of Charles Darwin*, vol. 19 (London: Pickering, 1988)

Darwin, Erasmus, Dr, *Erasmus Darwin: 'The Botanic Garden' 1791*, ed by Desmond King-Hele, facs. of 1791 edn (London: Scolar Press, 1973)

Darwin, Erasmus, Dr, *The Temple of Nature; or, The Origin of Society: a Poem, with Philosophical Notes, by Erasmus Darwin*, ed by E. D. Menston, facs. of 1803 edn (London: Scolar Press, 1973)

Darwin, Erasmus, Dr, *Zoonomia; or, The Laws of Organic Life*, 2 vols (London: J. Johnson, 1796)

Day, Thomas, 'The History of Little Jack' in *The Children's Miscellany* (London: Stockdale, 1804) pp. 1–54

Day, Thomas, *The History of Sandford and Merton* (London: J. Walker, 1808)

Descartes, René, *A Discourse on Method*, trans. by John Veitch (London: Dent, 1994)

Dickens, Charles, 'Frauds on the Fairies', *Household Words*, VIII. 184 (October 1853) pp. 97–100

Edinburgh Review or Critical Journal, XIII. 26 (1809) (Edinburgh: Archibald, Constable & Co, 1809)

Edwards, Sydenham, *Cynographia Britannica* (London: printed for the author, 1800)

Encyclopaedia Britannica; or a Dictionary of Arts, Sciences and Miscellaneous Literature, 3rd edn, 18 vols (Edinburgh: A. Bell and C. Macfarquhar, 1797)

Gentleman's Magazine, The, and Historical Chronicle, LXXXI. 2 (1811) (London: Nichols, Son and Bentley, 1811)

Gentleman's Magazine, The, and Historical Chronicle, LXXXIII. 1 (1813) (London: Nichols, Son and Bentley, 1813)

Goldsmith, Oliver, *The Citizen of the World* (London: Everyman Library, 1934)

Green, Joseph Henry, *Vital Dynamics: The Hunterian Oration Before the Royal College of Surgeons in London, 14 February 1840* (London: Pickering, 1840)

Hayley, William, *Ballads ... Founded on Anecdotes Relating to Animals; with prints designed and engraved by W. Blake* (Chichester: Richard Phillips, 1805)

Hegel, Georg Wilhelm Friedrich, *On Art, Religion, Philosophy: Introductory Lectures to the Realm of Absolute Spirit,* ed by J. Glenn Gray (New York: Harper and Row, 1970)

Hobbes, Thomas, *Leviathan; or, The Matter, Forme and Power of a Commonwealth, Ecclesiastical and Civil,* ed by Michael Oakeshott (Oxford: Basil Blackwell, 1957)

Hobhouse, John Cam, *Byron's Bulldog: The Letters of John Cam Hobhouse to Lord Byron,* ed by Peter W. Graham (Columbus: Ohio State University Press, 1984)

Hobhouse, John Cam, *Diary,* unpublished transcription by Peter Cochran from British Library Add. MSs. 56527

Hobhouse, John Cam, *Historical Illustrations to the Fourth Canto of* Childe Harold (London: Murray, 1818)

Hobhouse, John Cam, *John Cam Hobhouse, 'Imitations and Translations and The Wonders of a Week at Bath',* ed by Donald H. Reiman (New York: Garland, 1977)

Hogarth, William. *Hogarth's Graphic Works: First Complete Edition* (New Haven: Yale University Press, 1965)

Hogg, Thomas Jefferson, *Life of Percy Bysshe Shelley,* 2 vols (London: Edward Moxon, 1858)

Howard, Frederick, Earl of Carlisle, *Poems* (London: J. Ridley, 1773)

Hugo, Victor, *Préface de 'Cromwell' de Victor Hugo,* ed by Edmond Wahl (Oxford: Clarendon Press, 1932)

Hunt, Leigh, *The Autobiography of Leigh Hunt,* ed by J. E. Morpurgo (London: Cresset Press, 1949)

Hunt, Leigh, *Essays by Leigh Hunt,* ed by Arthur Symons (London: Walter Scott, 1888)

Hunt, Leigh, *The Examiner,* 592 (2 May 1819) (London: John Hunt, 1819)

Jeffrey, Francis, *Jeffrey's Criticism: A Selection,* ed by Peter F. Morgan (Edinburgh: Scottish Academic Press, 1983)

Jemmat, Catherine, *Miscellanies in Prose and Verse* (London: printed for the author, 1766)

Johnson, Samuel, *A Dictionary of the English Language,* facs. of Knapton edn, 1755 (London: Times, 1983)

Johnson, Samuel, *Johnson on Shakespeare: Essays and Notes Selected and Set Forth with an Introduction,* ed by Sir Walter Raleigh (London: Frowde, 1908)

Johnson, Samuel, *Lives of the English Poets,* ed by John Wain (London: Dent, 1975)

Journal of the House of Lords, vol. XLVII (1809–10)

Keats, John, *Keats, Poetical Works,* ed by H. W. Garrod (London: Oxford University Press, 1959, repr. 1967)

Keats, John, *The Letters of John Keats 1814–1821,* ed by Hyder Edward Rollins, 2 vols (Cambridge, Mass.: Harvard University Press, 1958)

Kendall, Augustus, *Keeper's Travels in Search of his Master* (London: E. Newbery, 1798)

Kilner, Dorothy, *The Life and Perambulations of a Mouse,* 2 vols (London: John Marshall, 1785)

Lamarck, Jean-Baptiste Monet, chevalier de, *Philosophie zoologique,* 2 vols (Paris, 1809)

Lamarck, Jean-Baptiste Monet, chevalier de, *Histoire naturelle des animaux sans vertèbres,* 7 vols (Paris, 1815–22)

Lamb, Charles, and Mary Anne Lamb, *The Letters of Charles and Mary Anne Lamb,* ed by Edwin W. Marrs Jr, 3 vols (Ithaca: Cornell University Press, 1976)

Lambe, William, *Additional Reports on the Effects of a Peculiar Regimen in Cases of Cancer, Scrofula, Consumption, Asthma, and other Chronic Diseases* (London: J. Mawman, 1815)

Lawrence, John, *A Philosophical and Practical Treatise on Horses; And on the Moral Duties of Man towards the Brute Creation,* 2 vols (London: 1796–98)

Lesage, Alain-René, *Le Diable Boiteux: texte de la deuxième édition avec les variantes de l'édition originale et du remaniement de 1726*, ed by Roger Laufer (Paris: Mouton, 1970)

Linnaeus, Carolus, 'The Oeconomy of Nature', in *Miscellaneous Tracts Relating to Natural History, Husbandry and Physick*, trans. (from *Amoenitates Academicae* by Carolus Linnaeus) by Benjamin Stillingfleet (London, 1749)

Locke, John, *Some Thoughts Concerning Education*, ed by F. W. Garforth (London: Heinemann, 1964)

London Magazine, The, X (2 October 1824) (London: Baldwin Cradock, 1824)

Macaulay, Thomas Babington, Lord, *The History of England from the Accession of James II*, 3 vols (London: J. M. Dent & Sons, 1906, repr. 1934)

Maitland, Henry, 'Marlow's Tragical History of the Life and Death of Doctor Faustus', in *Blackwood's Edinburgh Magazine* 1 (1817) pp. 393–4.

Malthus, Thomas, *On the Principle of Population* (London: J. Johnston, 1803)

Marlowe, Christopher, *Doctor Faustus*, ed by John D. Jump (London: Methuen, 1962)

Marvell, Andrew, *Andrew Marvell*, ed by Frank Kermode and Keith Walker (Oxford: Oxford University Press, 1994)

Medwin, Thomas, *Medwin's Conversations of Lord Byron* ed by Ernest J. Lovell Jr (Princeton: Princeton University Press, 1966)

Monboddo, James Burnet, Lord, *The Origin and Progress of Language* (Edinburgh: J. Balfour and T. Cadell, 1774)

Monboddo, James Burnet, Lord, *Antient Metaphysics, Volume Third, Containing the History and Philosophy of Men* (London: T. Cadell, 1784)

Montaigne, Michel, seigneur de, *Essays of Michel de Montaigne*, trans. by Charles Cotton, 3 vols (London: G. Bell and Sons, 1913)

Montesquieu, Charles de Secondat, baron de, *The Spirit of the Laws*, trans. and ed by Anne M. Cohler, Basia Carolyn Miller and Harold Samuel Stone (Cambridge: Cambridge University Press, 1989)

Moore, Thomas, *Letters and Journals of Lord Byron with Notices of his Life*, 2 vols (London: John Murray, 1830–31)

Moore, Thomas, *The Letters of Thomas Moore*, ed by Wilfred S. Dowden, 2 vols (Oxford: Oxford University Press, 1964)

Newton, John Frank, 'The Return to Nature; or, A Defence of the Vegetable Regimen, with some account of an experiment made during the last three or four years in the author's family,' *The Pamphleteer*, XIX. 38 (1821) pp. 497–530; XX. 39 (1822) pp. 97–118, and XX. 40 (1822) pp. 411–29

Oken, Lorenz, *Abriss der Naturphilosophie* (Göttingen, 1805)

Oken, Lorenz, *Die Zeugung* (Bamberg u. Wirzburg, 1805)

Oswald, John, *The Cry of Nature; An Appeal to Mercy and Justice on behalf of the Persecuted Animals* (London: J. Johnson, 1791)

Ovid, *Metamorphoses*, trans. by A. D. Melville, with introduction and notes by E. J. Kenney (Oxford: Oxford University Press, 1986)

Paine, Thomas, *Thomas Paine: Representative Selections*, ed by Harry Hayden Clark (New York: American Book Co., 1944)

Paley, William, *Natural Theology; or, Evidences of the Existence and Attributes of the Deity, Collected from the Appearances of Nature* (London: Faulder, 1803)

Parliamentary Debates from the Year 1803 to the Present Time, vol. XIV (London: Hansard, 1812)

Parliamentary Debates from the Year 1803 to the Present Time, vol. XVI (London: Hansard, 1812)

Parliamentary Debates from the Year 1803 to the Present Time, n.s., vol. VII (London: Hansard, 1825)

Parliamentary History of England, from the Earliest Period to the Year 1803, vol. XXXV (London: Hansard, 1819)

Parliamentary History of England, from the Earliest Period to the Year 1803, vol. XXXVI (London: Hansard, 1820)

Parry, William, *The Last Days of Lord Byron* (London: Knight and Lacey, 1825)

Peacock, Thomas Love, *Melincourt*, ed by Richard Garnett (London: Dent, 1891)

Pigot, Elizabeth, 'The Wonderful History of Lord Byron & His Dog' (unpublished manuscript booklet, 1807, in the Harry Hansom Research Center, University of Texas at Austin)

Piozzi, Hester Thrale, *Dr Johnson by Mrs Thrale: the 'Anecdotes' of Mrs Piozzi in their Original Form* ed by Richard Ingrams (London: Chatto & Windus, 1984)

Pliny, *Pliny: Natural History with an English Translation in Ten Volumes*, II, trans. by H. Rackham (London: Heinemann, 1942)

Plutarch, 'Beasts are rational', and 'On the Eating of Flesh', in *Plutarch's Moralia*, XII, trans. by Harold Cherniss and William C. Helmbold (London: Heinemann, 1957)

Pope, Alexander, *Poetical Works*, ed by Herbert Davis, 1st edn (1966) repr. with new introduction by Pat Rogers (Oxford: Oxford University Press, 1978)

Pope, Alexander, *The Correspondence of Alexander Pope*, ed by George Sherburn, 5 vols (Oxford: Clarendon Press, 1956)

Prior, Matthew, *The Literary Works of Matthew Prior*, ed by H. Bunker Wright and Monroe K. Spears (Oxford: Clarendon Press, 1959)

Quarterly Review, XVII (London: John Murray, 1817)

Quarterly Review, LI (London: John Murray, 1834)

Ritson, Joseph, *An Essay on Abstinence from Animal Food, as a Moral Duty* (London: Richard Phillips, 1802)

Robinson, Henry Crabb, *Henry Crabb Robinson on Books and their Writers*, ed by Edith J. Morley, 3 vols (London: Dent, 1938)

Rochester, John Wilmot, Earl of, *John Wilmot, Earl of Rochester: The Complete Works*, ed by Frank H. Ellis (London: Penguin, 1994)

Rousseau, Jean-Jacques, *Discourse on the Origins and Foundations of Inequality among Men* (Harmondsworth: Penguin, 1984)

Rousseau, Jean-Jacques, *Emile; or, Treatise on Education*, trans. by William H. Payne (London: Edward Arnold, 1901)

Rousseau, Jean-Jacques, *The Social Contract and Discourses*, trans. by G. D. H. Cole (London: Dent, 1973)

Saint-Hilaire, Isidore Geoffroy, *Histoire générale et particulière des anomalies de l'organisation chez l'homme et les animaux ... ou, traité de tératologie*, 3 vols (Paris, 1832–27)

Satirist, The, II (June 1808) (London: Samuel Tipper, 1808)

Seward, Anna, *The Poetical Works of Anna Seward, with Extracts from Her Literary Correspondence*, ed by Walter Scott, 3 vols (Edinburgh, 1810)

Shelley, Percy Bysshe, *Shelley, Poetical Works*, ed by Thomas Hutchinson (Oxford: Oxford University Press, 1970)

Shelley, Percy Bysshe, *Shelley's Prose; or, The Trumpet of a Prophecy*, ed by David Lee Clark (Albuquerque: University of New Mexico Press, 1954)

Shelley, Percy Bysshe, *The Complete Poetical Works of Percy Bysshe Shelley*, ed by Neville Rogers, vol. I, 1802–13 and vol. II, 1814–17 (Oxford: Clarendon Press, 1974–75)

Shelley, Percy Bysshe, *The Letters of Percy Bysshe Shelley*, ed by F. L. Jones, 2 vols (Oxford: Oxford University Press, 1964)

Shelley, Percy Bysshe, *The Prose Works of Percy Bysshe Shelley*, vol. I, ed by E. B. Murray (Oxford: Clarendon Press, 1993)

Smart, Christopher, *The Poetical Works of Christopher Smart*, vol. I, *Jubilate Agno*, ed by Karina Williamson (Oxford: Clarendon Press, 1980)

Southey, Robert, *A Choice of Southey's Verse*, ed by Geoffrey Grigson (London: Faber and Faber, 1970)

Southey, Robert, *Poems of Robert Southey*, ed by Maurice H. Fitzgerald, (London: Henry Frowde, Oxford University Press, 1909)

Spenser, Edmund, *The Works of Edmund Spenser: A Variorum Edition,* ed by Edwin Greenlaw, Charles Grosvenor Osgood and Frederick Morgan Padelford (Baltimore: Johns Hopkins Press, 1932)

Sterne, Laurence, *A Sentimental Journey through France and Italy by Mr Yorick,* ed by Gardner J. Stout Jr (Berkeley: University of California Press, 1967)

Swift, Jonathan, *The Correspondence of Jonathan Swift,* ed by Harold Williams, 5 vols (Oxford: Clarendon Press, 1962–65)

Swift, Jonathan, *Gulliver's Travels,* ed by Louis A. Landa (London: Methuen, 1965)

Taylor, Thomas, *A Vindication of the Rights of Brutes by Thomas Taylor,* ed by Louise Schutz Boas, facs. of 1792 edn (Gainesville, Florida: Scholars' Facsimiles & Reprints, 1966)

Tennyson, Alfred, Lord, *The Poems of Tennyson in Three Volumes: Second Edition Incorporating the Trinity College Manuscripts,* ed by Christopher Ricks (Harlow: Longman, 1969)

Thomson, James, *The Seasons: James Thomson,* ed by James Sambrook (Oxford: Clarendon Press, 1981)

Trimmer [Sarah], ed., *The Guardian of Education,* 5 vols (London: Hatchard, 1802–06)

Trimmer, [Sarah], *The History of the Robins, with Twenty-four Illustrations from Drawings by Harrison Weir,* first published as *Fabulous Histories,* 1786 (London: Griffith and Farran, n.d.)

Tryon, Thomas, *Pythagoras His Mystic Philosophy Revived; or, The Mystery of Dreams Unfolded* (London: Thomas Salisbury, 1691)

Voltaire, François-Marie Arouet, *The Complete Works of Voltaire,* vol. XXXV, *Dictionnaire philosophique* ed by Christiane Mervaud and others (Oxford: Voltaire Foundation, 1994)

Wilberforce, Robert I. and Samuel Wilberforce, *The Life of William Wilberforce,* 2 vols (London: John Murray, 1838)

Wollstonecraft, Mary, *Original Stories from Real Life 1791,* facs. of 2nd. edn, (Oxford: Woodstock Books, 1990)

Wollstonecraft, Mary, *The Works of Mary Wollstonecraft,* ed by Janet Todd and Marilyn Butler, 7 vols (London: William Pickering, 1989)

Wordsworth, Christopher, *Memoirs of William Wordsworth,* 2 vols (London: Ticknor, Reed and Fields, 1851)

Wordsworth, Dorothy, *Journals of Dorothy Wordsworth* ed by Mary Moorman (London: Oxford University Press, 1971)

Wordsworth, William, *Home at Grasmere, Part First, Book First, of* The Recluse *by William Wordsworth,* ed by Beth Darlington (Ithaca, New York: Cornell University Press, 1977)

Wordsworth, William, *The Prelude: 1799, 1805, 1850,* ed by Jonathan Wordsworth, M. H. Abrams and Stephen Gill (New York: Norton, 1979)

Wordsworth, William, *The Prose Works of William Wordsworth,* ed by W. J. B. Owen and Jane Worthington Smyser, 3 vols (Oxford: Clarendon Press, 1974)

Wordsworth, William, *William Wordsworth: Poetical Works,* ed by Ernest de Selincourt and Helen Darbishire, 5 vols (Oxford: Clarendon Press, 1940–49)

Wordsworth, William, and Dorothy Wordsworth, *The Letters of William and Dorothy Wordsworth,* ed by Ernest de Selincourt, 2nd edn, vol. I, *The Early Years 1787–1805,* rev. by Chester L. Shaver (Oxford: Clarendon Press, 1967)

Wordsworth, William, and Dorothy Wordsworth, *The Letters of William and Dorothy Wordsworth: The Middle Years, part one , 1806–1811,* ed by Ernest de Selincourt, 2nd edn, rev. by Mary Moorman (Oxford: Clarendon Press, 1969)

Wordsworth, William, and Dorothy Wordsworth, *The Letters of William and Dorothy Wordsworth: The Middle Years, part two: 1812–1820,* ed by Ernest de Selincourt, 2nd edn, rev. by Mary Moorman and Alan G. Hill (Oxford: Clarendon Press, 1970)

Wordsworth, William, and Dorothy Wordsworth, *The Letters of William and Dorothy Wordsworth: The Later Years, part one, 1821–8,* ed by Alan G. Hill, 2nd edn (Oxford: Clarendon Press, 1978)

Secondary sources cited

Baker, Steve, *Picturing the Beast: Animals, Identity and Representation* (Manchester: Manchester University Press, 1993)

Bakhtin, Mikhail Mikhailovich, *Rabelais and his World,* trans. by Helene Iswolsky (Bloomington: Indiana University Press, 1984).

Barrell, John, 'The Uses of Dorothy: "The Language of the Sense" in "Tintern Abbey"', in *New Casebooks: Wordsworth,* ed by John Williams (Basingstoke: Macmillan, 1993) pp. 142–71

Barron, Jeremy Hugh, and Arthur Crisp, 'Illness and Creativity: Byron's Appetites, James Joyce's Gut, and Melba's Meals and Mésalliances,' and 'Ambivalence towards Fatness and its Origins', *BMJ* 7123 (20–27 December 1997) pp. 1697–1703.

Barton, Anne, 'Lord Byron and Trinity: A Bicentenary Portrait' in *Trinity Review,* 1988 (Cambridge: Trinity College, 1988) pp. 3–6

Bate, Jonathan, *The Song of the Earth* (London: Picador, 2000)

Bate, Jonathan, *Romantic Ecology: Wordsworth and the Environmental Tradition* (London: Routledge, 1991)

Beer, Gillian, *Darwin's Plots: Evolutionary Narrative in Darwin, George Eliot and Nineteenth-Century Fiction* (London: Routledge & Kegan Paul, 1983)

Beer, Gillian, '"The Face of Nature": Anthropomorphic Elements in the Language of *The Origin of Species',* in *Languages of Nature: Critical Essays on Science and Literature,* ed by Ludmilla J. Jordanova (London: Free Association Books, 1986) pp. 207–43

Berger, John, 'Why look at animals?' in *About Looking* (London: Writers and Readers, 1980) pp. 1–26

Berlin, Sir Isaiah, *The Roots of Romanticism,* ed by Henry Hardy (London: Chatto & Windus, 1999)

Bettelheim, Bruno, *The Uses of Enchantment: The Meaning and Importance of Fairy Tales* (London: Thames and Hudson, 1976)

Bewell, Alan, *Wordsworth and the Enlightenment: Nature, Man and Society in the Experimental Poetry* (New Haven: Yale University Press, 1989)

Bloom, Harold, *The Anxiety of Influence,* 2nd edn (London: Oxford University Press, 1997)

Blount, Margaret, *Animal Land, The Creatures of Children's Fiction* (London: Hutchinson, 1974)

Boas, George, *The Happy Beast in French Thought of the Seventeenth Century* (Baltimore: Johns Hopkins Press, 1909, repr. New York: Octagon Books, 1966)

Bottrall, Margaret, ed., *William Blake, Songs of Innocence and Experience: A Casebook* (London: Basingstoke, 1970)

Brantley, Richard E., *Wordsworth's 'Natural Methodism'* (New Haven: Yale University Press, 1975)

Brett, R. L. and A. R. Jones (eds), *'Lyrical Ballads': Wordsworth and Coleridge. The Text of the 1798 edition with the additional 1800 poems and the Prefaces edited with introduction, notes and appendices* (London and New York: Routledge, 1963 (2nd edn 1991, repr. 1993)).

Brewer, William D., *The Shelley-Byron Conversation* (Florida: University Press, 1994)

Brooke, John Hedley, *Natural Theology in Britain from Boyle to Paley* (Milton Keynes: Open University Press, 1974)

Brown, Margaret, '"The Firmest Friend" – Boatswain: Lord Byron's Dog,' in *Tails of the Famous* by Elizabeth Edwards and Margaret Brown (Bourne End, Buckinghamshire: Kensal, 1987) pp. 13–29

Bryant, Julius, *London's Country House Collections* (London: Scala Publications / English Heritage, 1993)

Buell, Lawrence, *The Environmental Imagination: Thoreau, Nature Writing, and the Formation of American Culture* (Cambridge, Mass.: Harvard University Press, 1995)

Bush, Clive, *The Dream of Reason: American Consciousness and Cultural Achievement from Independence to the Civil War* (London: Edward Arnold, 1977)

Butler, Samuel, *Evolution, Old and New; Or, the Theories of Buffon, Dr Erasmus Darwin, and Lamarck, as Compared with that of Charles Darwin* (London: Fifield, 1911)

Canovan, Margaret, 'Rousseau's Two Concepts of Citizenship', in *Women in Western Political Philosophy: Kant to Nietzsche,* ed by Ellen Kennedy and Susan Mendus (Brighton: Wheatsheaf, 1987) pp. 78–105

Chapple, J. A. V., *Science and Literature in the Nineteenth Century* (Basingstoke: Macmillan, 1986)

Christie, Will, 'Wordsworth and the Language of Nature', in *Wordsworth Circle,* XIV.1 (Winter 1983) pp. 40–47

Clark, Kenneth, *Animals and Men: Their Relationship as Reflected in Western Art from Prehistory to the Present Day* (London: Thames and Hudson, 1977)

Clark, Timothy and Mark Allen, 'Between Flippancy and Terror: Shelley's "Marianne's Dream"', in *Romanticism* I. 1 (1995) pp. 90–105

Clarke, M. L., *Paley: Evidences for the Man* (London: SPCK, 1974)

Coope, Rosalys, *Lord Byron's Newstead* (no publisher's details, 1988)

Curtis, Penelope, Matthew Craske, Jonathan Burt and Steve Baker, *Hounds in Leash: The Dog in 18th and 19th Century Sculpture* (Leeds: Henry Moore Institute, 2000).

Darnton, Robert, *The Great Cat Massacre and Other Episodes in French Cultural History* (London: Penguin, 1985)

Darton, F. J. Harvey, *Children's Books in England: Five Centuries of Social Life,* 3rd edn, rev. by Brian Alderson (Cambridge: Cambridge University Press, 1982)

de Almeida, Hermione, *Romantic Medicine and John Keats* (Oxford: Oxford University Press, 1991)

Desmond, Adrian, 'Artisan Resistance and Evolution in Britain, 1819–1848', in *Osiris,* 2nd series, III (1987) pp. 77–110

Desmond, Adrian, 'Lamarckism and Democracy: Corporations, Corruption and Comparative Anatomy in the 1830s' in *History, Humanity and Evolution, Essays for John C. Greene,* ed by James R. Moore (Cambridge: Cambridge University Press, 1989) pp. 99–130

Dictionary of National Biography ed by Leslie Stephen and Sidney Lee, 63 vols and supplement (London: Smith, Elder, 1885–1901)

Dobzhansky, Theodosius, *Mankind Evolving: The Evolution of the Human Species* (New Haven: Yale University Press, 1962)

Douglas, Mary, *Purity and Danger: An Analysis of the Concepts of Pollution and Taboo* (London: Routledge & Kegan Paul, 1966)

Encyclopaedia Britannica, ninth edition (London, A. & C. Black, 1881–8)

Franklin, Caroline, *Byron's Heroines* (Oxford: Clarendon Press, 1992)

Franklin, Caroline, 'Juan's Sea Changes: Class, Race and Gender in Byron's *Don Juan*' in *Don Juan* ed by Nigel Wood (Buckingham: Open University Press, 1993) pp. 56–89

Freud, Sigmund, *The Standard Edition of the Complete Psychological Works of Sigmund Freud,* ed by James Strachey, 24 vols (London: Hogarth Press and Institute of Psycho-Analysis, 1953–74)

Fudge, Erica, *Perceiving Animals: Humans and Beasts in Early Modern English Culture* (Basingstoke: Macmillan, 2000)

Fyfe, Aileen, 'The Reception of William Paley's *Natural Theology* in the University of Cambridge', *British Journal of the History of Science,* XXX (1997) pp. 321–35

Gelpi, Barbara Charlesworth, *Shelley's Goddess: Maternity, Language, Subjectivity* (New York: Oxford University Press, 1992)

Gilbert, Sandra M. and Susan Gubar, *The Madwoman in the Attic: The Woman Writer and the Nineteenth-century Literary Imagination* (New Haven: Yale University Press, 1979)

Glacken, Clarence J., *Traces on the Rhodian Shore: Nature and Culture in Western Thought from Ancient Times to the End of the Eighteenth Century* (Berkeley: University of California Press, 1967)

Grabo, Carl, *Shelley, a Newton among Poets: Shelley's Use of Science in* Prometheus Unbound (Chapel Hill: University of North Carolina Press, 1930)

Gradman, Barry, *Metamorphosis in Keats* (Brighton: Harvester Press, 1980)

Graham, Peter W., 'Byronic Darwinizing' in *Lord Byron: A Multidisciplinary Open Forum, proceedings of the 23rd International Byron Conference, Versailles, 26–30 June, 1997,* ed by Thérèse Tessier (n. p., 1999) pp. 125–34

Graham, Peter W., 'The Order and Disorder of Eating in Byron's *Don Juan*' in *Disorderly Eaters: Texts in Self-Empowerment,* ed by Lilian J. Furst and Peter W. Graham (Pennsylvania: University of Pennsylvania Press, 1992) pp. 113–24

Hawtree, Christopher, ed., *The Literary Companion to Dogs* (London: Sinclair-Stevenson, 1993)

Henson, Eithne, *'The Fictions of Romantick Chivalry': Samuel Johnson and Romance* (Rutherford: Fairleigh Dickinson University Press, 1992)

Holmes, Richard, *Shelley: The Pursuit* (London: Weidenfeld and Nicolson, 1974)

Ingold, Tim, ed., *What is an Animal?* (London: Hyman, 1988)

Jordanova, Ludmilla, 'Nature's Powers: A Reading of Lamarck's Distinction between Creation and Production' in *History, Humanity and Evolution: Essays for John C. Greene,* ed by James R. Moore (Cambridge: Cambridge University Press, 1989) pp. 71–98.

Kean, Hilda, *Animal Rights: Political and Social Change in Britain since 1800* (London: Reaktion, 1998)

Kelsall, Malcolm, *Byron's Politics* (Sussex: Harvester Press, 1987)

Killham, John, *Tennyson and 'The Princess': Reflections of an Age* (London: Athlone Press, 1958)

King-Hele, Desmond, *Erasmus Darwin and the Romantic Poets* (Basingstoke: Macmillan, 1986)

Lamb, Jeremy, *So Idle a Rogue: The Life and Death of Lord Rochester* (London: Allison and Busby, 1993)

Leeson, Robert, *Children's Books and Class Society: Past and Present* (London: Children's Rights Workshop, 1977)

Levere, Trevor H., *Poetry Realized in Nature: Samuel Taylor Coleridge and Early Nineteenth-century Science* (Cambridge: Cambridge University Press, 1981)

Levin, Susan M., *Dorothy Wordsworth & Romanticism* (London: Rutgers State University, 1987)

Levinson, Marjorie, *Keats's Life of Allegory: The Origins of a Style* (Oxford: Basil Blackwell, 1988, repr. 1990)

Levinson, Marjorie, *Wordsworth's Great Period Poems: Four Essays* (Cambridge: Cambridge University Press, 1986)

Lévi-Strauss, Claude, *The Savage Mind* (London: Weidenfeld and Nicolson, 1966)

Lévi-Strauss, Claude, *Totemism,* trans. by Rodney Needham (Boston: Beacon Press, 1962)

Lindop, Grevel, 'The Language of Nature and the Language of Poetry', *Wordsworth Circle* XX. 1 (Winter 1989) pp. 2–9

Liu, Alan, *Wordsworth: The Sense of History* (Stanford: Stanford University Press, 1989)

Lloyd-Jones, Ralph, '"Boatswain is Dead!" But When? A Mystery,' in *The Newstead Abbey Byron Society Newsletter,* XII (Winter, 1997) pp. 3–9

Lovell, Ernest J. Jr, *Byron: The Record of a Quest: Studies in a Poet's Concept and Treatment of Nature* (Austin: University of Texas Press, 1949)

Lovell, Ernest J. Jr, *His Very Self and Voice: Collected Conversations of Lord Byron* (New York: Macmillan, 1954)

Manning, Peter J., 'Childe Harold in the Marketplace: From Romaunt to Handbook', in *Modern Language Quarterly,* LII (1991) pp. 170–90

Marchand, Leslie A., *Byron: A Biography,* 3 vols (New York: Alfred A. Knopf, 1957)

Martin, Philip W., *Byron: A Poet before his Public* (Cambridge: Cambridge University Press, 1982

Martin, Philip W., 'Reading *Don Juan* with Bakhtin' in *Don Juan* ed by Nigel Wood (Buckingham: Open University Press, 1993) pp. 90–121

McGann, Jerome J., *The Poetics of Sensibility: A Revolution in Literary Style* (Oxford: Clarendon Press, 1996)

McGann, Jerome J., *The Romantic Ideology: A Critical Investigation* (Chicago: University of Chicago Press, 1983)

McIntosh, Robert P., *The Background of Ecology: Concept and Theory* (Cambridge: Cambridge University Press, 1985)

McNeil, Maureen, *Under the Banner of Science: Erasmus Darwin and his Age* (Manchester: Manchester University Press, 1987)

Meigs, Cornelia and others, *A Critical History of Children's Literature* (New York: Macmillan, 1953)

Mellinkoff, Ruth, *The Mark of Cain* (Berkeley: University of California Press, 1981)

Mellor, Anne K., *Romanticism and Gender* (New York: Routledge, 1993)

Midgley, Mary, *Beast and Man: The Roots of Human Nature* (London: Routledge, rev. edn, 1995)

Moore, Doris Langley, *Ada, Countess of Lovelace* (London: John Murray, 1977)

Moore, Doris Langley, *The Late Lord Byron: Posthumous Dramas* (London: John Murray, 1961)

Morton, Timothy, *Shelley and the Revolution in Taste: The Body and the Natural World* (Cambridge: Cambridge University Press, 1994)

Mulvey, Laura, 'Visual Pleasure and Narrative Cinema', in *Screen*, XVI. 3 (Autumn 1975) pp. 6–18.

Newlyn, Lucy, *Coleridge, Wordsworth and the Language of Allusion* (Oxford: Clarendon Press, 1986)

Nurmi, Martin K., 'Blake's Revisions of "The Tyger",' in *William Blake: Songs of Innocence and Experience: A Casebook,* ed by Margaret Bottrall (London: Macmillan, 1970) pp. 198–217

Paterson, Wilma, *Lord Byron's Relish: The Regency Cookery Book* (Glasgow: Dog and Bone, 1990)

Paulson, Ronald, 'Blake's Revolutionary Tiger', in *William Blake's* Songs of Innocence and of Experience, ed by Harold Bloom (New York: Chelsea House, 1987) pp. 123–32

Peach, Annette, 'Portraits of Byron' reprinted from *The Walpole Society*, LXII (2000)

Pendred, G. L., 'Mad, Bad and Dangerous Dogs,' in *Country Life*, 182. 10 (March 10, 1988) pp. 156–57

Penny, Nicholas B., 'Dead Dogs and Englishmen', in *The Connoisseur*, 192. 774 (August 1976) pp. 298–303

Pickering, Samuel F. Jr, *John Locke and Children's Books in Eighteenth-Century England* (Knoxville: University of Tennessee Press, 1981)

Pinkney, Tony, 'Romantic Ecology' in *A Companion to Romanticism* ed by Duncan Wu (Oxford: Blackwell, 1998) pp. 411–19

Piper, William Bowman, 'Gulliver's account of Houyhnhnmland as a Philosophical Treatise', in *The Genres of* Gulliver's Travels, ed by Frederick Smith (Newark: University of Delaware Press, 1990) pp. 179–202

Porter, Roy, *Doctor of Society: Thomas Beddoes and the Sick Trade in Late-Enlightenment England* (London and New York: Routledge, 1992)

Regan, Tom, 'Honey Dribbles down your Fur', in *Environmental Ethics: Philosophical and Policy Perspectives* ed by Philip P. Hanson (Burnaby BC: Institute for the Humanities/SFU Publications, 1986) pp. 99–113

Richardson, Alan, 'Romanticism and the Colonization of the Feminine' in *Romanticism and Feminism,* ed by Anne K. Mellor (Bloomington: Indiana University Press, 1988) pp. 13–25

Richardson, Alan, 'Wordsworth, Fairy Tales and the Politics of Children's Reading', in *Romanticism and Children's Literature in Nineteenth-Century England,* ed by James Holt McGavran Jr (Athens and London: University of Georgia Press, 1991) pp. 34–53

Righter, Anne, 'John Wilmot, Earl of Rochester' (Chatterton Lecture on an English Poet, read 18 January, 1967), *Proceedings of the British Academy,* LIII (1967) (Oxford: Oxford University Press, 1968) pp. 47–69

Ritvo, Harriet, *The Animal Estate: The English and Other Creatures in the Victorian Age* (Cambridge, Mass.: Harvard University Press, 1987)

Ritvo, Harriet, *The Platypus and the Mermaid and other Figments of the Classifying* Imagination (Cambridge, Mass.: Harvard University Press, 1997

Robinson, Charles E., *Shelley and Byron: The Snake and Eagle Wreathed in Fight* (Baltimore and London: Johns Hopkins University Press, 1976)

Roe, Nicholas, *The Politics of Nature: Wordsworth and some Contemporaries* (Basingstoke: Macmillan, 1992)

Rollins, Hyder Edward, ed., *The Keats Circle,* 2nd. edn, 2 vols (Cambridge Mass.: Harvard University Press, 1965)

Rosenfield, Leonora Cohen, *From Beast-Machine to Man-Machine: Animal Soul in French Letters from Descartes to La Mettrie* (New York: Octagon Books, 1940, repr. 1968)

Rudwick, Martin J., *Georges Cuvier, Fossil Bones, and Geological Catastrophes: New Translations and Interpretations of the Primary Texts* (Chicago: University of Chicago Press, 1997)

Rutherford, Andrew, ed., *Byron: The Critical Heritage* (London: Routledge & Kegan Paul, 1970)

Ryder, Richard D., *Animal Revolution: Changing Attitudes towards Speciesism* (Oxford: Basil Blackwell, 1989)

Sadler, Sir Michael, *Thomas Day: An English Disciple of Rousseau,* the Rede Lecture, 1928 (Cambridge: Cambridge University Press, 1928)

Scholtmeijer, Marion, *Animal Victims in Modern Fiction: From Sanctity to Sacrifice* (Toronto: University of Toronto Press, 1993)

Sedgwick, Eve Kosofsky, *Between Men: English Literature and Male Homosocial Desire* (New York: Columbia University Press, 1985)

Sells, A. Lytton, *Animal Poetry in French and English Literature and the Greek Tradition* (London: Thames and Hudson, 1957)

Serpell, James, *In the Company of Animals: A Study of Human-Animal Relationships* (Oxford: Basil Blackwell, 1986)

Shaver, Chester L., and Alice C. Shaver, *Wordsworth's Library: A Catalogue* (New York: Garland, 1979)

Singer, Peter, *The Expanding Circle: Ethics and Sociobiology* (Oxford: Clarendon Press, 1981)

Smith, Christine, 'Lord Byron: A Dog's Best Friend', in *Kennel Gazette,* August 1991, pp. 12–17.

Smith, Hillas, *Keats and Medicine* (Newport: Cross Publishing, 1995)

Smith, R. J., *The Gothic Bequest: Mediaeval Institutions in British Thought, 1688–1863* (Cambridge: Cambridge University Press, 1987)

Sorabji, Richard, *Animal Minds and Human Morals: The Origins of the Western Debate* (Ithaca: Cornell University Press, 1993)

Stafford, Fiona, 'Fingal and the Fallen Angels: Macpherson, Milton and Romantic Titanism': lecture at *Influence and Intertextuality* conference, University of Bristol, 24 May 1997

St Clair, William, 'William Godwin as Children's Bookseller' in *Children and their Books: A Celebration of the Work of Iona and Peter Opie,* ed by Gillian Avery and Julia Briggs (Oxford: Clarendon Press, 1989) pp. 165–80

Steiner, George, *After Babel: Aspects of Language and Translation* (Oxford: Oxford University Press, 1992)

Stevenson, Lionel, *Darwin among the Poets* (New York: Russell, 1932 repr. 1963)

Stryker, Lloyd Paul, *For the Defence: Thomas Erskine, One of the Most Enlightened Men of his Times, 1750–1823* (London and New York: Staples Press, 1949)

Summerfield, Geoffrey, *Fantasy and Reason: Children's Literature in the Eighteenth Century* (London: Methuen, 1984)

Tapper, Richard, 'Animality, humanity, morality, society', in *What is an Animal?* ed by Tim Ingold (London: Hyman, 1988)

Taylor, Anya, *Coleridge's Defense of the Human* (Columbus: Ohio State University Press, 1986)

Tester, Keith, *Animals and Society: The Humanity of Animal Rights* (London: Routledge, 1991)

Thomas, Keith, *Man and the Natural World: Changing Attitudes in England 1500–1800* (London: Allen Lane, 1983)

Thrupp, 'The Antient Mariner and the Modern Sportsman' (London: James Martin, 1881)

Turner, E. S., *All Heaven in a Rage* (London: Michael Joseph, 1964)

Turner, James, *Reckoning with the Beast: Animals, Pain, and Humanity in the Victorian Mind* (Baltimore and London: John Hopkins University Press, 1980)

Warner, Marina, *No Go the Bogeyman* (London: Chatto & Windus, 1998)

Watson, Jeanie, 'Coleridge and the Fairy Tale Controversy,' in *Romanticism and Children's Literature in Nineteenth-Century England,* ed by James Holt McGavran Jr (Athens and London: University of Georgia Press, 1991) pp. 14–33

Whalley, George, 'The Mariner and the Albatross' in *Coleridge:* The Ancient Mariner *and Other Poems: A Casebook* ed by Alun R. Jones and William Tydeman (London: Macmillan, 6th repr. 1990) pp. 160–83

White, Newman Ivey, *Shelley,* 2 vols (New York: Alfred Knopf, 1940)

Williams, John, *New Casebooks: Wordsworth* (Basingstoke: Macmillan, 1993)

Williams, Raymond, *The Country and the City* (London: Hogarth,1973)

Williams, Raymond, *Marxism and Literature* (Oxford: Oxford University Press, 1977)

Wilson, Douglas B., *The Romantic Dream: Wordsworth and the Poetic of the Unconscious* (Lincoln: University of Nebraska Press, 1993)

Wilson Knight, G., 'The Two Eternities,' in *Modern Critical Views: George Gordon, Lord Byron* ed by Harold Bloom (New York: Chelsea House Publishers, 1986) pp. 39–52

Wolfson, Susan J., '"Their She Condition": Cross-dressing and the Politics of Gender in *Don Juan',* in *Journal of English Literary History,* LIV (1986) pp. 585–617.

Woodring, Carl R., *Politics in the Poetry of Coleridge* (Madison: University of Wisconsin Press, 1961)

Woof, Robert, Stephen Hebron and Claire Tomalin, *Hyenas in Petticoats: Mary Wollstonecraft and Mary Shelley* (Kendal: Wordsworth Trust, 1997)

Woolf, Virginia, *Flush: a Biography* (London: Hogarth Press, 1947)

Wordsworth, Jonathan, *William Wordsworth: The Borders of Vision* (Oxford: Clarendon Press, 1982)

Worthington, Jane, *Wordsworth's Reading of Roman Prose* (Newhaven: Yale University Press, 1946)

Worster, Donald, *Nature's Economy: A History of Ecological Ideas,* 2nd edn (Cambridge: Cambridge University Press, 1977, repr. 1994)

Wu, Duncan (ed.), *Romanticism: An Anthology* (Oxford: Blackwell, 1994, repr. 1999)

Wu, Duncan, *Wordsworth's Reading 1800–1815* (Cambridge: Cambridge University Press, 1995)

Wyatt, John, *Wordsworth and the Geologists* (Cambridge: Cambridge University Press, 1995)

Wylie, Ian, *Young Coleridge and the Philosophers of Nature* (Oxford: Clarendon Press, 1989)

Young, Robert M., *Darwin's Metaphor: Nature's Place in Victorian Culture* (Cambridge: Cambridge University Press, 1985)

Index